Apocalypse When?

Calculating How Long the Human Race Will Survive

Willard Wells

Apocalypse When?

Calculating How Long the Human Race Will Survive

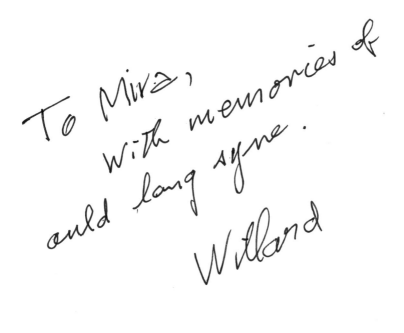

To Mira,
with memories of
auld lang syne.

Willard

 Springer

Published in association with
Praxis Publishing
Chichester, UK

Dr Willard Wells
Consultant, Applied Physics, Communications
Chief Scientist Emeritus, L-3 Photonics
Carlsbad
California
USA

SPRINGER–PRAXIS BOOKS IN POPULAR SCIENCE
SUBJECT *ADVISORY EDITOR*: Stephen Webb, B.Sc., Ph.D.

ISBN 978-0-387-09836-4 Springer Berlin Heidelberg New York

Springer is part of Springer-Science + Business Media (springer.com)

Library of Congress Control Number: 2008942384

Cover design: Jim Wilkie
Project copy editor: Laura Booth
Typesetting: OPS Ltd, Gt Yarmouth, Norfolk, UK

Printed in Germany on acid-free paper

Contents

Preface ix

Acknowledgments xiii

List of figures xv

List of pictures xvii

List of tables xix

List of algebraic symbols xxi

List of abbreviations and acronyms xxiii

About the author xxv

Introduction 1

1   Formulation 11
     1.1   Multiple hazard rates 14
     1.2   Probability theory: a quick review 18
     1.3   Changing hazard rates 20
     1.4   Posterior probability 21
     1.5   Principle of indifference 25
     1.6   Cumulative risk 34

2   Confirmation 39
     2.1   Bayes' theory 39
     2.2   Statistics of business firms 41
     2.3   Statistics of stage productions 45

2.4   Longevity rank   52
2.5   Univariate summary   53

3   Double jeopardy   57
3.1   A paradox   57
3.2   Formulation   58
3.3   An example   61
3.4   Logic diagram   64
3.5   Future research   65

4   Human survivability   67
4.1   Formulation   69
4.2   Hazardous development   73
4.3   Predictor formulation   81
4.4   Survivability of civilization   84
4.5   Survivability of the human race   86
4.6   Summary and current hazard rates   89
4.7   Biases   91

5   Apocalypse how?   93
5.1   Scenarios for extinction   94
5.2   Wild cards   105
5.3   Overrated natural hazards   112
5.4   Triage   114
5.5   Reflections on this study   116
5.6   Prospects for a safer world   117
5.7   Systemic strengths and weaknesses   120
5.8   Second chance?   123
5.9   Survival habitat   125

APPENDICES

A   Survival formula derived from hazard rates   129
A.1   Variable hazard rate   131
A.2   Unknown vulnerability   132

B   Posterior survivability   135

C   Infinite mean duration   137

D   Survival predictor from Bayes' theory   139
D.1   Advance knowledge   142
D.2   Noninformative prior probability   145

E   Stage productions running on specified dates   147

F   Extinction rates of prehistoric taxa   159

G   Disaggregated mortality   163

H   Stage productions with dual cum-risks   167

I   Overall plan for survivability calculation   187

J   Multiple hazards   189

K   Cum-risks for man-made hazards   193

L   Statistical weights for hazards   197

M   Extinction thwarted by civilization's collapse   199

N   Initial hazard rates   203

References   205

Index   209

To Judith who encouraged me throughout this project
and
To the memory of Richard Feynman who taught me to think
unconventionally

Preface

Starting about 1950, our world changed forever. Global population from then until 2000 multiplied by 2.5 and technology surged forward at an unprecedented pace. For some people these developments are a cause for great concern. Sir Martin Rees, England's present Astronomer Royal, thinks this pace may cause a fatal error within a century or two. He has wagered $1000 that a single act of bio-error or bio-terror will kill a million people before 2020. Bill Joy, co-founder of Sun Microsystems, fears that machines will overpower us using cybertechnology that he helped create. Stephen Hawking, renowned physicist and author of *A Brief History of Time*, thinks we must colonize outer space in order to ensure survival of our species.

We are already taking measures to protect ourselves from conspicuous hazards such as global warming and genetic engineering. Hence, the fatal one will most likely be some bizarre combination of events that circumvents normal safeguards. For example, mutant phytoplankton may spread across the oceans and poison the air with toxic gasses. Or a demented trillionaire may imagine God's command to exterminate humanity at all cost, and with his vast resources he may succeed. The chances of these particular hazards are minuscule, but there are hundreds more like them, so the overall risk is considerable.

Is there a way to quantify these concerns? We could build survival habitats if a numerical assessment justified the expense. The best numerical data to address this question would be survival statistics for humanoid species throughout our galaxy, especially those on Earth-like planets undergoing a technology surge. However, those data are scarce, so what else can we do?

The obvious uninspired approach would make a huge numerical model of our entire world and run thousands of simulations of our future, each with slightly different inputs and random events. Statistics of the outcomes would then indicate major threats and the probability of surviving them. Computer programs already simulate gigantic physical systems, for example, world climate. The Club of Rome has made and operated a huge numerical model of the world's economy. An essential missing piece is a numerical model for human behavior, but even that may be available in fifty years. If current trends continue, computer capability will have increased many thousandfold by then!

Would such simulations yield credible estimates of the ultimate risk? Not likely. History does not provide disasters of this magnitude to use as test cases. If it did, we could pretend to live in an earlier time and then run the program to "predict" events comparable to those that have actually happened. Such test runs would help us adjust parameters in the prediction program and correct flaws. In particular, the program must randomly inject rare events, tipping points, and extraordinary genius with some frequency, but that frequency must be adjusted. Otherwise, the rarities occur too often or too seldom. We might try to calibrate by using records of lesser historical disasters. However, few if any extant reports of those early disasters include enough details to initialize a simulation. Thes obstacles may be insurmuntable.

If we somehow get a credible simulation, it may indicate an urgent need for harsh reforms that offend almost everybody: levy heavy taxes on consumption of natural resources, especially fossil fuels; impose compulsory birth control; and so on. The public reaction to this report would itself make an interesting subject for a world simulation. One can imagine the repercussions: Special interests hire scientists to ridicule the world model. The simulation team is branded as alarmists trying to inflate their importance. The public is helpless because the simulation's flow chart alone is too complex for any individual to grasp. Even dedicated independent review teams would struggle long and hard with the vast number of algorithms and statistics.

#   #   #

The subject of this book is a more immediate and practical approach to survivability, a simple analytic model that transcends the quagmire of details. It is so simple that you can keep your daytime job and still find time to challenge it. Perhaps you can revise it using different sources of statistics.

The formulation relies on two measures of past survival: one for exposure to natural hazards, and the other for man-made hazards. The first measure is comforting. Humankind has survived natural hazards for 2000 centuries. After such long exposure, we can surely expect at least another 20 centuries, merely 1% longer. By contrast, the second measure is worrisome. Our exposure to serious man-made hazards has been a scanty half-century, which means that lack of experience leaves us vulnerable. New hazards appear faster than we can safely adapt to recently established ones.

#   #   #

A scholar normally has a duty to use extensive data in an effort to achieve maximum accuracy. However, human survival is an ideologically sensitive subject, and the burden of so much data might discourage constructive criticism and future revisions by others. Besides, the more analysts tinker with data, the more they must wrestle with personal prejudices that (consciously or not) could bias the results, and the greater the risk that one of their sources will be discredited, or that their analysis has (or appears to have) an ideological slant. In our curious case the reader's acceptance is more important than a modest increase in accuracy. Consequently, my analysis is simple and the input data minimal.

'Tis better to be approximate and credible than to be exact and ignored.

This tradeoff runs contrary to my training as a physicist, but that's a whole different scene. One gains credibility in physics when another laboratory independently repeats and confirms one's results. The trouble is that people won't be repeating the human extinction experiment. And by the time a forecast is verified, it is too late. Hence, in this peculiar instance, I believe that a formulation that emphasizes credibility over accuracy is justified and prudent. It is also user-friendly because a revision takes only a few sessions at a desktop computer.

The text has been written in such a way that the reader unfamiliar with math should have little trouble in following the arguments. The mathematically sophisticated reader can find a full development in the appendices.

Acknowledgments

Henri Hodara introduced me to J. Richard Gott's survival predictor, which got me hooked on this project. He also gave much useful advice, but best of all, he coined the term "Gott-erdämmerung."

Stephen Webb reviewed my manuscript and made many useful suggestions.

Richard Puetter reviewed my manuscript and found two logical errors.

Debates with Allen Ashley clarified my interpretation of Gott's predictor.

José Mata sent me his preprint about business survival and guided me to other business data.

Dale Novina edited my manuscript for style and legibility.

Trevor Lipscombe suggested the title. He and Michael Fisher made helpful suggestions.

For updates and additional statistics, go to

www.ApocalypseWhen.com

Figures

1   Longevity of Portuguese business firms 12
2   Exponential decay 15
3   Survival of radioactive ficticium 16
4   Timelines for Gott's survival predictor 27
5   Probability G of future F given age A 30
6   Logic diagram, part 1 of 2 40
7   Timeline for a process of known duration 41
8   Longevity of Canadian service firms 42
9   Same data with modified scales 43
10   Longevity of Canadian goods producers 44
11   Longevity of manufacturing firms 44
12   Stage productions, choice of entities 46
13   Compositions in London, 1950–51 47
14   Anomalous decay in 1900 and 1901 48
15   Survival of compositions, 1920 to 1924 49
16   Survival of shows open on specified dates 51
17   Scatter diagram for London shows 63
18   Logic diagram, part 2 of 2 65
19   U.S. patents granted yearly 76
20   Pages in *Nature* magazine, yearly 77
21   Papers in natural sciences and engineering 77
22   Gross world product, sum of GDPs 78
23   Statistical weights for exponents 83
24   Civilization's survivability 85
25   Basic species survivability 86
26   Survivability in terms of projected population-time 88
27   Summary plots of survivability 89
28   Choices for philanthropy 116
29   Extinction insurance 123

Pictures

Entrance to the Svalbard Global Seed Vault.............................. 2
http://en.wikipedia.org/wiki/Image:Global-Seed-Vault_0844_inngansparti_kunst_ F_Mari_Tefre.jpg
(last accessed January 3, 2009)
Wikimedia Commons; thanks to Mari Tefre

Underground Isotope Production Facility............................... 16
http://www.osti.gov/bridge/servlets/purl/840202-2GdRZ6/native/840202.pdf
(last accessed January 3, 2009)
Los Alamos National Laboratory

Hot cell for processing isotopes.................................... 17
http://www.northstarnm.com/Technology.htm
(last accessed January 3, 2009)
Courtesy of Argonne National Laboratory

Cumulative risk meter.. 20

Human serial numbers: Adam and Eve by Albrecht Dürer, 16th century, spliced to
personal photos.. 33

Spaceman Jorj pondering his oil supply 35

Haymarket Theatre 50
London Wikimedia Commons; thanks to K. B. Thompson

London's Wyndham's Theatre in 1900................................ 51
Wikimedia Commons; copyright expired

Zipf and his law.. 53

Earth bomb.. 91

McMurdo Station, Antarctica 94
http://www.passcal.nmt.edu/~bob/passcal/antarctica/ant03.htm
(last accessed October 2008)
Thanks to Robert Greschke, photographer

Volcanic Deccan Traps in India.................................... 99
http://en.wikipedia.org/wiki/Image:Kille_Rajgad_from_Pabe_Ghat.jpg
(last accessed October 2008)
Wikimedia Commons

Electric stork 102
http://en.wikipedia.org/wiki/Image:Stork_nest_on_power_mast.jpg
(last accessed January 3, 2009)
Wikimedia Commons; thanks to Germán Meyer, photographer

Deforestation in Amazonia 107
http://en.wikipedia.org/wiki/Image:Amazonie_deforestation.jpg
(last accessed January 3, 2009)
Source NASA

Comet impacts on Jupiter 114
http://www.solarviews.com/eng/impact.htm
(last accessed January 3, 2009)
Courtesy of NASA IRTF Comet Science Team

Native copper 124
Wikimedia Commons; thanks to Jonathan Zander

Red tide . . . . back cover
Original source and permission from NIWA, New Zealand
(National Institute of Water & Atmospheric Research)

Four Horsemen of the Apocalypse . . . . front cover
Albrecht Dürer from 16th century

Tables

1	Behavior of the mean future as sample size grows	23
2	Benford's law for frequency of leading digits	33
3	Standard deviations from multivariate theory	64
4	Summary quantities	64
5	Dispassionate look at social values	70
6	Conversions between dates and population-time	75
7	Proxies for hazardous development	78
8	Relative risks during three years	79
9	Hazards during 1970 compared to 2005	80
10	Initial hazard rates	90

Algebraic symbols

(appendices include some ad hoc notations)

Upper-case Roman alphabet

A	age of entity
B	bad fraction of humanity
C	constant
E	world economy
F	future time
G	Gott's survival predictor: Prob(future | age)
H	inverse of G: Prob(age | duration)
J	shortest time for hazard to act, related to gestation
K	shortest cum-risk for hazard to act, complement to
L	lost stage productions
M	pop-time of modern mankind since 1530 AD
N	number (count), various
O	quantity of oil on Planet Qwimp
P	past time
Q	prior probability of duration (at time of birth)
R	survival rank, $R = 1$ being longest duration
S	counted performances
T	entity's duration, past + future (mnemonic: term)
U	measure of hazardous economic and technical development, haz-dev
V	counted shots at shooting gallery
W	weighted probability of q, given as $W(q)$
X	population-time accrued since the dawn of humanity (mnemonic: extended)
Y	turnstile count at shooting gallery
Z	cumulative risk (hazard), cum-risk

Lower-case Roman

a	subscript for age
c	subscript for civilization
f	subscript for future
m	subscript for man-made
n	subscript for natural
p	subscript for past
p	world population
q	exponent for a Gott factor, usually paired with $1 - q$
r	exponent for a second Gott factor
s	subscript for our species

Greek

$\alpha, \beta, \ldots, \omega$	series of exponents totaling 1.0
λ	hazard rate
Λ	initial (now) hazard rate to our species, used with subscripts s and c; see above
μ	feedback exponent for haz-dev U
ω	exponent for converting pop-time (X) to haz-dev (U) and cum-risk (Z)

Abbreviations and acronyms

AIDS	Acquired Immune Deficiency Syndrome
BP	Billion People
BPC	Billion People-Centuries
BSE	Bovine Spongiform Encephalopathy
GDP	Gross Domestic Product
GA	Genetic Algorithm
GSP	Gott's Survival Predictor
GWP	Gross World Product
NASA	National Aeronautics and Space Administration
NSE	Natural Sciences and Engineering
pdf	probability density function
SES	Secret Eugenics Society
SEC	Securities and Exchange Commission
TPY	Tera-People-Years (tera = trillion)

About the author

Dr. Wells holds a Ph.D. in physics with a minor in mathematics from the California Institute of Technology (Caltech), where his mentor was Prof. Richard Feynman, Nobel Laureate. Wells worked at Caltech's Jet Propulsion Laboratory where he analyzed the unexpected tumbling of Explorer I satellite. This led to an improved design for later Explorers. He was co-inventor of a mechanism that stops a satellite's spin after it attains orbit. This device was used in several missions during the early years of the space program. He then founded and led the Quantum Electronics Group.

Dr. Wells spent most of his career as chief scientist of Tetra Tech Inc., an engineering firm doing research and development under contract to clients, mostly military. He is now Chief Scientist Emeritus of L-3 Photonics in Carlsbad, California and a member of San Diego Independent Scholars.

Introduction

It's a poor sort of memory that only works backwards.

—Lewis Carroll

The great mathematician John von Neumann (1903–57) once famously said, "The ever accelerating progress of technology ... gives the appearance of approaching some essential singularity [an abnormal mathematical point] in the history of the race beyond which human affairs, as we know them, could not continue." Another math professor, Vernor Vinge, well known for his science fiction, picked up the concept. He began lecturing about the *Singularity* during the 1980s and published a paper about the concept in 1993 [1].

The ideas caught the imagination of non-scientists who founded a secular moral philosophy in 1991. They call themselves "Singularitarians" and look forward to a technological singularity including superhuman intelligence. They believe the Singularity is both possible and desirable, and they support its early arrival. A related philosophy is called *transhumanism*, followers of which believe in a *post-human* future that merges people and technology via bioengineering, cybernetics, nanotechnologies, and the like.

By contrast, a number of renowned scientists think that the advent of the Singularity is a time of great danger. The pace of technological innovation is accelerating so rapidly that new waves of progress appear faster than we can safely acclimate to other recently established ones. Sooner or later we shall make a colossal mistake that leads to apocalypse. The killer need not be a single well-known hazard like global warming. Scientists who are alert to this sort of threat are watching indicators for signs of danger and will likely warn us before the hazard becomes critical. Instead, the killer will likely be something we overlook, perhaps a complex coincidence of events that blindsides us. Or it may be a devious scheme that a

Entrance to the Svalbard Global Seed Vault.

misanthrope conceives, somebody with the mentality of the hackers who create computer viruses.

Sir Martin Rees, England's current Astronomer Royal, thinks the 21st century may be our last [2], the odds being about 50–50. He has wagered $1000 that a single act of bio-error or bio-terror will kill a million people by 2020.

Perhaps the most gravely worried is Bill Joy, co-founder of Sun Microsystems and inventor of the Java computer language. He contributed much to the cyber-technology that he now fears [3]: "There are certain technologies so terrible that you must say no. We have to stop some research. It's one strike and you're out." Again, "We are dealing now with technologies that are so transformatively powerful that they threaten our species. Where do we stop? By becoming robots or going extinct?"

Physicist Stephen Hawking, renowned author of *A Brief History of Time*, has joined the chorus: "Life on Earth is at the ever-increasing risk of being wiped out by a disaster, such as sudden global warming, nuclear war, a genetically engineered virus, or other dangers we have not yet thought of." To my knowledge no eminent scientist has publicly objected to any of these concerns.

In June 2006 the five Scandinavian prime ministers gathered at Spitsbergen, a Norwegian island in the far arctic, part of the Svalbard archipelago. The occasion was the groundbreaking ceremony to begin construction of a $5-million doomsday vault that will store crop seeds in case of a global calamity. Again, prominent people have a vague but widespread perception of serious danger.

Optimists disagree including the late economist Julian Simon [4] and the late novelist Dr. Michael Crichton [5]. Crichton has reviewed past predictions of doom and, from their failure, he has inferred that all such predictions are invalid. This is clearly a *non sequitur* since all prophecies of doom fail—except the last.

All these pundits, both optimists and pessimists, rely on qualitative reasoning that is endlessly debatable. By contrast, the arguments presented here are quantitative. Numerical results are also debatable, but to a lesser extent. Doubts converge on two parameters. I hope that readers will use this model, and with modest effort and different sources of data compute their own survival figures. Detailed results will vary, but the most robust conclusions will prevail.

#   #   #

Some folks question whether it is valid to apply an impersonal mathematical formula to human survival, given that we are intelligent conscious creatures with ideals and spiritual qualities. To answer this objection, we can represent humanity by microcosms for which actual survival statistics are readily available. In particular, business firms and theatrical productions share important qualities with humanity. Those two microcosms plus humanity comprise three entities with the following attributes in common:

- All three consist of people striving for the entity's survival.
- All are exposed to many diverse hazards.
- All are aggregates of individuals, each of whom can be replaced while the entity remains intact.
- Within each entity the individuals act from mixed motives that balance group interests against personal ones.
- None of the three entities (our species, business firms, stage productions) has a cutoff age, a maximum it cannot exceed.

However, there is one relevant difference. People in businesses and in theater work together for the common good and develop a sense of teamwork and group consciousness. This does not extend to our species as a whole, which is too vast and amorphous for such feelings to take hold. With regard to cooperation, our species as a whole is the *least* "human" of the three entities. Therefore we might expect our species to conform *better* to a dispassionate, indifferent formula than the "more human" microcosms do.

Shakespeare would approve of stage plays as a microcosm for human survival: "All the world's a stage, and all the men and women merely players. They have their exits and their entrances" (*As You Like It*, 2:7).

#   #   #

Technology, industry, and population, all feed on one another to produce a dangerous level of development. Bulldozers raze jungles to expand agriculture. Machinery plows the land, plants crops, and harvests them. The abundant food then promotes population, more people to make more machines to produce still bigger crops. This so-called *positive feedback* runs faster and faster. Agrochemistry produces fertilizer and insecticides, which make more food followed by more people, who consume more fertilizer and insecticides and deplete natural resources. People work in sweatshops and make more computers, which free our time to build more industry and invent more technology. The processing power of computers doubles every few years. Everything will accelerate until civilization hits some hard physical limit and breaks down. The human race may be part of that breakdown.

All these survival risks are far too complex and chaotic to analyze directly. We cannot attempt to extrapolate current trends. That would be valid only briefly until our world encounters a *tipping point*. After that, trends change suddenly, drastically and unpredictably. The study of such instability is known as *chaos* or *complexity theory*. The best-known paradigm of chaos is the proverbial butterfly in Brazil that can cause (or prevent) a tornado in Texas if it decides to flit from one twig to the next. The butterfly's tiny slipstream encounters an unstable airflow and alters its motion out of proportion to its original size. These secondary currents in turn alter bigger instabilities, and so on until the train of events grows to hurricane size.

Such bizarre causality applies to each single instance. However, in the statistics of many tornadoes and many butterflies, the number of times a butterfly causes a tornado offsets the number of times another butterfly prevents one. Thus, for big samples, the number of tornadoes and butterflies are unrelated on average, as you would expect. In other words, statistical numbers conform to smooth predictable formulas despite the chaos in particular instances.

Imagine living in 1900 and trying to anticipate nuclear winter, cobalt bombs, and global warming. Such predictions were impossible then and are just as impossible now. Therefore, our formulation for studying long-term human survival is not based on numerical simulations of future events or on any detailed risk analysis. Instead, it transcends these imponderables by relying on humanity's two-part track record for survival: one part is past exposure to natural hazards, and the other is past exposure to man-made hazards.

A rough analogy may help here. Suppose you measure the overall properties of a big machine—inputs and outputs such as power consumption, power delivered, temperature, pressure, and entropy. Then you apply the laws of thermodynamics to the machine as a whole and deduce something about its performance without analyzing all the forces on individual gears and wheels. In principle the detailed analysis would convey more information, but in practice it may be inaccessible, too costly, impractical and/or prone to error. Likewise we put aside the details of world simulation and rely instead on humanity's dual histories of survival.

Our track record for surviving natural hazards is comforting. Humankind has survived them for 2000 centuries. After such long exposure, we can surely expect to survive another 20 centuries, only 1% longer, other things being equal. But other things are not equal. Our exposure to serious man-made hazards has lasted a scanty half century. New hazards appear faster than we can safely acclimate to established hazards. Our job is to balance this huge disparity in exposure and arrive at a best estimate for our species' longevity.

Had we known our species' age in 1900, only the first survival history would apply, the one for natural hazards. In that case, as we shall see, the formula from this book reduces to a simpler formula first discovered by astrophysicist J. Richard Gott [6]. Imagine that you were living in 1900 and discovered Gott's formula. Using it, you would have predicted a 90% chance of survival for at least another 22,000 years— nothing to be concerned about.

For comparison, our ancestor, *Homo erectus*, lasted 1.6 million years—8 times longer than our current age. Our cousins, the Neanderthals, lasted 300,000 years—

50% longer. By all these measures our species is still adolescent, much too young to worry about survival issues. However, those relatives did not make artifacts capable of mass destruction, nor did we in 1900. At that time the risks were all natural events such as asteroid strike or climate change. No man-made hazard was then powerful enough to consummate self-extinction, nor did anyone expect such hazards in the future.

Then something extraordinary happened: world population soared, and so did the pace of technological innovation. Starting about 1950 we acquired the means to cause or at least contribute to self-extinction. Nature might start an epidemic, but modern air travel could spread it to remote places with unprecedented speed. Consumption of fossil fuel altered the atmosphere. Human extinction became a feasible scientific project, perhaps by genetic engineering or robotics. And nobody knows what is next. We have slipped out of the safe stable equilibrium of past centuries and are hurdling toward an unknown Singularity. We have the power to forestall hazards if we anticipate them in time, but when they bombard us too rapidly, anticipation will fail someday when the big hazard hits. To quote James Thurber, "Progress was alright; it only went on too long" [7].

#   #   #

As we shall see, numerical results show that the risk of extinction is currently 3% per decade, while the risk of a lesser apocalyptic event, the collapse of civilization, is triple that, about 10% per decade, or 1% per year. The magnitude of these risks is comparable to ordinary perils that insurance companies underwrite. (Perhaps they should offer extinction insurance!) Both risks are proportional to world population. Every time the population doubles, so does our risk. Unless the population plummets soon, a *near-extinction event will likely occur during the lifetime of today's infants.*

The collapse of civilization will be almost as deadly as extinction. However, for those who do survive, the aftermath will be a very safe time. Sparse population will protect outlying villages from epidemics. The biosphere will recover from human degradation, although perhaps with a changed climate and extinction of many species.

Ironically, humanity's long-term survival *requires* a worldwide cataclysm. Without it we are like the dinosaurs. They seem to have expired because the big bolide (meteoric fireball) hit first without warning. Had there been precursors, near-extinctions, many species might have adapted well enough to survive the big one. Since humankind will likely suffer a near-extinction event; the probability of long-term survival is encouragingly high, roughly 70%.

(Incidentally, this high probability bears on Fermi's paradox: the universe apparently has a great many Earth-like planets, so why have we not seen evidence of intelligent life? As Stephen Webb has mentioned [8], one explanation says that extinction is the normal result when humanoids reach our present phase of high-tech development. However, some of those exohumanoid species should have high survivability similar to our own and thus live to explore the galaxy and leave footprints. The apparent absence of intelligent life probably has some other explanation.)

# # #

Think of self-extinction as a job to be accomplished, like digging a ditch. Suppose it takes 6 man-hours of manual labor to dig a ditch, then it takes 1 man 6 hours, or 2 men 3 hours, or 3 men 2 hours. However, in our case the number of people involved is not the size of a work crew but rather the entire world population. And the time involved is not the duration of a job but rather the entire lives of the people. Therefore, in place of labor expressed as man-hours, we have population-time (pop-time for short) expressed as billions of people centuries, abbreviated BPC. Thus pop-time is a measure of total human life worldwide.

(Incidentally, the grand total for all past human pop-time is about 1.7 trillion people-years, calculated by adding the world's estimated population for every year since the dawn of our species. The uncertainty of prehistoric population matters little because their population was so small.)

Labor to do a job often cannot be accurately predicted. Ditch diggers going out on a job may not know what they will encounter. The earth may be soft soil, rocky, or hardpan. Thus, in advance they can only estimate labor with varying degrees of confidence. They may estimate that a certain ditch will require 4 hours with 30% confidence, or 8 hours at 90%. Likewise, we can only estimate the amount of pop-time to consummate extinction at various degrees of confidence. The half-life of civilization, defined as 50% confidence, will last about 8.6 BPC. This will accrue in about a century if the population is 8.6 billion. The half-life of the human race is about 30 BPC. However, this will accrue extremely slowly after civilization collapses, which will probably happen first.

If hi-tech civilization thrives long enough, there will come a window of opportunity for sending intelligent life into outer space, ultimately to colonize the solar system and possibly the galaxy. That window opens when we acquire the requisite technology, and it closes when we exhaust essential resources. It is difficult to estimate the time and probability of this opening and closing, but at least our species will most likely be alive to exploit it if and when it does occur.

Our descendants may colonize the solar system and perhaps the galaxy either with biological humans or with some sort of conscious artifacts that we regard as our intellectual descendants (cyborgs, androids, whatever). When these colonial descendants achieve independence from Earth's resources, they will then be almost immune to extinction. Loss of any one habitat will not threaten the others.

# # #

Any formula for human survival will surely conflict with somebody's worldview, thereby inviting controversy. It is critical, therefore, that the reasoning be as thorough and credible as possible. We shall proceed cautiously with many examples using four very different approaches or viewpoints. All four converge on approximately the same formula. Four approaches may seem excessive, but the math is vague in some places (fuzzy, as mathematicians often say), and the strengths of one argument compensate weaknesses in another.

One of these approaches has an abstract quality that seems almost unreal because it mentions nothing about hazards, risk rates, or causality. Instead, it is all about the moment we observe the entity in question and when that moment occurs during its life span. Nevertheless, it produces almost the same formula as the other approaches and provides valuable insight for generalizing the basic theory. This approach stems from the so-called *Doomsday Argument* [9, 10], which began about 1983. This argument comprises the main historical background for our topic, so let us digress briefly to examine it now.

The main players in the Doomsday Argument are about a dozen scholars, mostly philosophers and physicists, and especially astrophysicists. In its usual form the argument goes like this: Starting from the dawn of our species, the total number of people who have ever lived is about 70 billion. If the world ends next month, those of us alive now will be among a small fraction at the very end of human history. On the other hand, if humanity lasts for several millennia and colonizes the galaxy, then trillions will live. In that case we will be among a tiny fraction at the very beginning of human history. Both extremes are statistically unlikely; it is much more probable that we now live sometime in the big middle of human history.

Let us choose a time in the "big middle" to use as a reference. The most natural choice is the exact middle, the median, defined as the time when the number of future lives equals the number of past lives, about 70 billion. To estimate this future, let us use the worldwide birthrate. It is now about 160 million annually; suppose that it stabilizes during this century at 200 million/year. Our median future is then

$$70B \div 0.20B/yr = 350 \text{ years.}$$

Similar reasoning, which seems equally plausible, gives a very different result. In 1993 astrophysicist J. Richard Gott III independently made the same argument [6] except that he based his estimates on time instead of counted lives. He noted that humanity began about 2000 centuries ago, and that we are probably not at the extreme beginning or the extreme end of humanity's duration. On this basis (time), our median future is another 2000 centuries, which is drastically different from the first estimate, 3.5 centuries, the ratio being 570. As we shall see, this huge discrepancy is soluble by working from the various approaches mentioned above, each of which is clear in one aspect and occasionally murky in another.

Meanwhile, the Doomsday Argument continues. Its basic quantity or *independent variable*—such as time or counted lives—has not been decided. Participants recognize that its choice is crucial, but they have not settled on a criterion. Moreover, whatever quantity they choose, they must also decide what lives to count. Should they include future cyborgs? Androids? All conscious entities in our galaxy? Who (or what) comprises the so-called *reference class*? Scholars realize that hi-tech risks are important; hence some say that the reference class should be something like computer owners, or perhaps people who understand the Doomsday Argument. If this choice were valid, extinction would be imminent. Whenever we discuss the survivability of some entity, that entity must be defined in such a way that its beginning is known and its end is clearly defined.

So, although the Doomsday Argument contains a core of truth, it leaves two ill-defined concepts: the independent variable and the reference class. Using inputs from two other approaches, these problems are explained and resolved in Sections 1.3 and 1.6 below, which then make the Doomsday Argument quantitative and useful. In this formulation the reference class is always the biological species *Homo sapiens* and risks are defined in ways consistent with that choice.

Professor Gott substantiated his survival formula using data from 44 stage productions advertised in *The New Yorker* magazine on May 27, 1993. His example led me to use stage productions as one of the two microcosms that represent humanity. However, his analysis was sketchy [11]. To my knowledge, no other doomsday scholar has used *any* statistic from the real world. Data that appear in Sections 2.2, 2.3 and 3.3 below fill this gaping omission.

One of the most active scholars in the doomsday group is Canadian philosopher John Leslie. In the introduction to his book [12], *The End of the World*, Leslie states on page 3:

> "The doomsday argument aims to show that we ought to feel *some* reluctance to accept any theory which made us very exceptionally early among all humans who would ever have been born [emphasis in original]. The sheer fact that such a theory made us very exceptionally early would at least strengthen any reasons we had for rejecting it. Just how much would it strengthen them? The answer would depend on just how strong the competing reasons were—the reasons for thinking that the human race would survive for many more centuries, perhaps colonizing the whole of its galaxy. The competition between reasons might even be modeled mathematically."

J. Richard Gott III

Dr. Gott is a professor of astrophysics at Princeton University. He is best known as the author of *Time Travel in Einstein's Universe: The Physical Possibilities of Travel through Time*, 2002, Houghton Mifflin Books.

He is also well known for his involvement in the Doomsday Argument, which he discovered independently. He claims that his epiphany came when he visited the Berlin Wall as a tourist in 1969 and wondered how long it would stand. Knowing very little of the geopolitical issues, which were too complex anyhow, he looked for and found a principle that would estimate its future based on a single datum, its age. He relates his viewpoint to the Copernican principle, the idea that humans are not privileged observers of the universe, as he is not a privileged observer of the Berlin Wall. (Copernicus was the Polish astronomer who in the 16th century showed that Earth is not the center of the Solar System, but rather just a geometrically unprivileged planet.)

Prof. Gott received the President's Award for Distinguished Teaching in recognition of his work with the National Westinghouse and Intel Science Talent Search, a competition for high school students.

That last sentence succinctly summarizes the project described in this book. Recall the examples of independent variables discussed above, counted lives and time. Imagine that these were the relevant ones for human survival. Counted lives says that extinction is nigh, while time says that there is no problem, and it is too soon to worry. These are examples of Leslie's "competing reasons". If counted lives and time were the relevant variables, the technique used here would consider the relative importance of the two and produce numbers for survival times at various levels of confidence.

\#   \#   \#

You will see equations in this book, but they are relatively simple and their key features are described in words. So even if you know little mathematics, you can still follow the arguments. For mathematically sophisticated readers, full derivations and details are given in the appendixes. Five chapters follow:

- Chapter 1 derives the survival predictor and its variants from four viewpoints.
- Chapter 2 discusses and fortifies the formulation with statistics and additional context.
- Chapter 3 shows how to manage divergent risks, essentially Leslie's "competing reasons". In particular, the risk of natural disaster is spread evenly in time—a volcano is as likely to erupt one decade as another. By contrast, man-made hazards such as genetic engineering are concentrated in modern times and are still accelerating. So how best can we choose a hazard rate between these extremes?
- Chapter 4 adapts the predictor to human survival and presents the main numerical results. In addition to survival of the human race it includes prospects for civilization.
- Finally, Chapter 5 relaxes the quantitative discipline and includes conjecture, opinion, general principles for survival, and a list of serious hazards. With sufficient awareness we can hope to beat the odds and make our species one of the more durable survivors in our galaxy. Perhaps we can hang on long enough for a viable colony to escape somewhere or at least take refuge in an artificial habitat.

1

Formulation

*The value of a formalism lies not only in the range
of problems to which it can be successfully applied,
but equally in the degree to which it encourages physical
intuition in guessing the solutions of intractable problems.*

—Sir Alfred Brian Pippard

Let us begin by examining the most basic formula for an entity's survival. The equation gives its decreasing probability of survival starting at birth. For a stage production that would be curtain time on opening night. Later we shall adjust that formula to give the entity's survival probabiliy starting at a later time when it is observed alive.

In Chapter 2 we substantiate this simple equation using many sets of statistics, but for now let us examine a single set to illustrate the concept. This example is a set of data compiled by Mata and Portugal, which lists the longevity of domestic business firms in Portugal. Their tabular data appear as points in Figure 1, which shows the fraction Q of firms surviving after a duration T. (The word *term* can serve as a mnemonic for T, but it is ambiguous compared to *duration*.)

Mata and Portugal had access to annual reports submitted to the Portuguese Ministry of Employment. Portuguese law requires everyone hiring labor to report business statistics, so the data include all sizes of firms, even the smallest. Therefore, bias in the data due to the firms' size and importance is minimal. The authors used data for more than 100,000 firms from 1982 to 1992 and distilled them to eight summary data in their Table 6 under the heading "domestic survival rate".

If we assume that the business climate is stable, meaning it is the same now as it was during the lives of the firms in this set, then we can interpret Q as the probability that a new firm will remain open for business beyond time T.

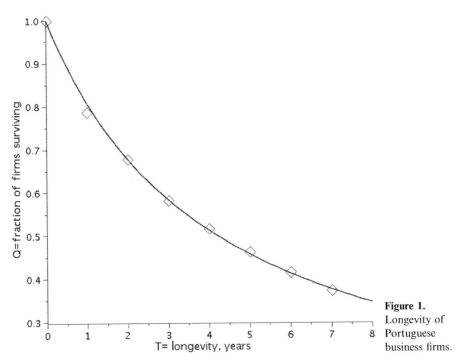

Figure 1. Longevity of Portuguese business firms.

The smooth curve in Figure 1 has the formula

$$Q = \frac{J}{J + T} = \frac{1}{1 + T/J} \qquad (1)$$

where J is a curve-fitting parameter. In this case $J = 4.2$ years happens to fit the points almost exactly.

By itself the fit in Equation 1 is not statistically significant. However, more statistics in Sections 2.2 and 2.3 agree, and so do two theories in Sections 1.1 and 1.5. The combined weight of all this support makes the formula compelling. Of the two expressions in Equation 1, the first is more convenient for most mathematics, while the second emphasizes that the ratio T/J is what counts, not T and J individually.

Many later formulas in this treatise are mere modifications and generalizations of this main equation. It applies to our two microcosms and by inference to the human race. However, parameter J differs from one entity to another. It differs not only among broad classes such as stage production and business firms, but also among categories such as service firms and manufacturing.

It is tempting to interpret J as a gestation period. Since T is measured from birth, negative T represents the prenatal period during which hazards can cause miscarriage. This idea extends back to $T = -J$, at which time the formula breaks down. (It gives $Q = J \div$ zero, which is of course nonsense: the operation of dividing by zero is undefined.) This breakdown might be acceptable, however, if we interpret $T = -J$ as the inchoate moment of conception.

However, the real definition of J is something related but different. It is the time during which the fastest hazard dispatches its victims (as Appendix A explains). For example, stage productions do not expire as the curtain rises on opening night except in the most extraordinary circumstance. (Perhaps a meteor strikes the theater at that moment.) Similarly, business firms do not expire while the owner is unlocking the door on her first day of business. At some time longer than these absurd extremes, there is a practical time J in which the faster hazards act. It is related to the actual gestation period, because any hazard that acts in less time dispatches vulnerable entities prior to birth, and is therefore never observed in the statistical ensemble. The next section gives an example. However, nothing prevents J from being longer than gestation.

As already mentioned, for the Portuguese businesses in Figure 1 the best fit to the curve is $J = 4.2$ years. My friend Henri Hodara commented on this number since he and three partners founded a successful technology firm in 1966. Henri said that 4.2 years is much too long; preparations for their firm lasted only about 6 months. However, there is another factor: one of the four partners had previously founded a company in the same market, and all of them had worked together for years. Thus the founders' track record for survival was already in place before the corporation legally existed. Clearly, this makes the concept of business gestation rather fuzzy. However, 4.2 years may well be reasonable for a business formed from scratch by people who have never worked together. In any case, let us continue to use the word *gestation* as a nickname for J because a fully descriptive name for the concept would require too many words.

If we put $T = J$ in Equation 1 then we find $Q = \frac{1}{2}$. In other words, J is also the median survival time, and indeed a thumb rule for business states that the first 5 years (close enough to 4.2) are critical. A firm that passes that milestone will probably succeed.

Note that there is no limit to the duration of a business firm. Accordingly, Equation 1 does not go to zero at a finite cutoff time. This equation therefore cannot apply to entities having an inflexible age limit, such as the human body. Nor does it apply to anything with a known constant hazard rate, defined as a probability of expiring per unit time. A classic example of the latter is a radioisotope that decays exponentially with a known half-life, which we discuss in the next section. Instead, Equation 1 applies best when the risks are numerous and diverse, and/or the hazard rates are completely unknown, as explained in the following section, the first of the theoretical methods mentioned earlier. We shall see later that a second theory produces similar results. More survival statistics in Chapter 2 also substantiate Equation 1. Taken together, these approaches make a compelling case for this equation.

As mentioned in the introduction, we shall approach the formulation of humanity's survival from four viewpoints:

- survival statistics for business firms and stage productions (first example above, many more in Section 2.2)
- probability theory based on random hazard rates (Section 1.1)
- probability theory based on our history of survival (Section 1.4)
- Bayesian theory (Section 2.1)

The first approach substantiates the main formula using actual survival statistics for microcosms. However, it offers no theoretical insight that would help modify and extend the theory to dual threats, as needed to compute human survival.

The second approach, theory based on hazard rates, shows how to deal with variable risks, in particular the accelerating technologies that threaten humankind.

The third approach is a revision of the Doomsday Argument, which was briefly summarized in the introduction. Despite its eccentricity, the argument produces almost the same formula for mankind's survival as the other approaches and gives valuable insight for generalizing the basic theory.

The fourth approach begins with a trivial formula for the probability of an entity's age if we already know its duration. It then uses Bayes' theorem to invert this formula and obtain what we really want, the probability of duration given age.

In the following section we proceed to examine one of the theoretical approaches.

1.1   MULTIPLE HAZARD RATES

A radioactive atom has a constant hazard rate, a fixed chance of expiring per unit time regardless of its age. Its familiar decay law is the exponential curve in Figure 2. A bulk sample of a radioisotope decays by half after a time appropriately known as its half-life. Atoms that survive the first half-life learned nothing from their experience, nor have they a will to live. Therefore, their hazard rate remains the same during the next half-life, and so half of the remainder decays, leaving a quarter. This halving continues until the last atom vanishes. The dotted lines on Figure 2 mark the succession of halves, 50%, 25%, 12.5% ...

Besides radioisotopes, many other things decay exponentially. They include electric charge on a capacitor that leaks through a fixed resistance, red light penetrating blue water, and so forth. In some cases, the decay represents survival of certain objects, for example molecules in a metastable state, or particles in a beam penetrating a gas. These objects have no absolute age limit because their parts do not wear out. Instead, the number of survivors dwindles gradually due to a constant risk rate as shown in Figure 2. A mundane example is the supply of Harry Potter books at San Diego's public library. (They kindly gave me their statistics.) Although books do eventually disintegrate, the dominant hazard is borrower's failure to return them, a fixed probability per loan.

Some critics mentioned exponential decay as a counterexample to the generalization of Equation 1 (beyond Portuguese businesses). Two of these critics were P. Buch [14] and E. Sober [15], but they did not understand the underlying assumptions. Equation 1 pertains to *lack* of knowledge. If you know the half-life, it is

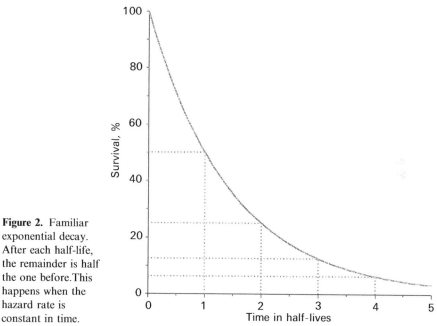

Figure 2. Familiar exponential decay. After each half-life, the remainder is half the one before. This happens when the hazard rate is constant in time.

not supposed to apply. It does apply if you do *not* know the half-life or if you have a mixed sample with many different half-lives (see Figure 3 below). This kind of probability is a tricky subject: it depends not only on the physical properties of the system or process but also on what the observer knows about it. You may know or suspect that a die is loaded but not know which face it favors. From your perspective the prior probabilities all remain 1/6 until you observe a few rolls and reassess. Your overall probability blurs the distinction between indeterminacy (the individual roll of the die) and lack of information (the die's inner structure).

Consider a sample of the radioactive element ficticium, freshly prepared in a nuclear reactor. It has four isotopes in equal abundance with decay rates 1, 3, 5, and 7 per week. (The decay rate equals $0.693 \div$ half-life.) Figure 3 shows the fraction of ficticium surviving after time T measured in weeks. Dotted curves show the decay of individual isotopes. The bold solid curve shows their average, the fraction of all ficticium atoms surviving in the mixed batch. The bold dashed curve is a plot of our survival formula, Equation 1, with $J = 1/6$ week $= 1.2$ days. Over the time interval shown here, our formula is a fair approximation to the actual survival shown in the bold solid curve.

Ficticium has more unstable isotopes with greater decay rates (shorter half-lives), but we are scarcely aware of them because they have mostly decayed by the time the specimen is extracted from the nuclear reactor, purified, and delivered to the laboratory, which is about a day in this example. If the delivery had taken less time, say an

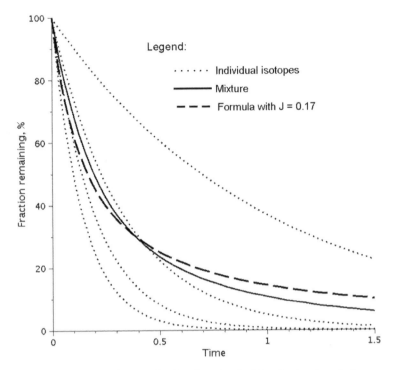

Figure 3. Survival of a sample of radioactive ficticium having four isotopes.

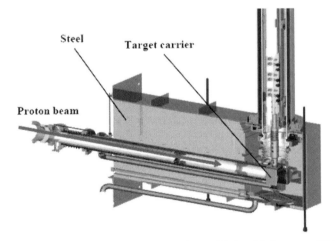

Underground Isotope Production Facility at Los Alamos National Laboratory. This schematic view shows a proton beam striking a target to make radioisotopes. The target carrier, a steel box, is surrounded by concrete. A lift in the vertical shaft carries the target into a hot cell above ground where an operator can manipulate the fresh isotopes without exposure to radiation.

Hot cell for processing
isotopes. Note robotic
manipulators and
chemical mixing agents.

hour, we would then see these short-lived isotopes decaying. They would make the
initial slope in Figure 3 much sharper and would fit Equation 1 with a much smaller
value of J, about 1 hour. Therefore, we can think of delivery time as a gestation
period, which imposes a maximum on the observable hazard rate. This is one
explanation for the parameter J, although other limitations on risk can make J
greater.

The curves in Figure 3 represent a hierarchy of uncertainties. If you know an
atom's isotope and the half-life of that isotope, then you know all that is physically
possible to know about that atom's survival prospect. And yet you cannot predict its
survival time with certainty because its decay is an indeterminate process. All you can
know about the single isotope is its probability of survival, which appears in the
appropriate dotted curve.

The solid curve represents the next level of uncertainty. It tells us the survival
probability of a single atom drawn at random from the batch of ficticium. You know
the isotopic abundances and the half-life of each, but you do not know which isotope
you drew. The same solid curve also represents the surviving fraction of the original
mixture (as we shall see in the next section).

Now consider another case, in which the isotopes are separated into four bulk
samples. You draw one of these samples at random, not knowing which one it is.
However, once you have observed the sample's decay for a while you can determine
which isotope it is by measuring its decay rate. The remaining fraction does not
follow the solid curve but rather one of the dotted ones for that pure isotope.
Nonetheless, *prior to observing its decay*, the solid curve was still your best estimate
of the decay, as proven in Section 1.2 below.

Next, suppose you first draw a random isotope and then draw an atom from that
sample. Now the probability that you drew a particular atom is the same as though

the whole supply had been physically mixed, and so the atom's survival probability reverts to that case, the solid curve again.

Finally, consider the case in which you know nothing about a sample except that it is radioactive and something about its preparation time, which lets you estimate J. Now the dashed curve given by Equation 1 is your best initial estimate of survival. All these different kinds of uncertainty could be given names and carefully identified as we encounter them, but that seems pedantic. Instead, let us not belabor these distinctions, but rather let the interpretation of each case follow naturally from its context. Appendix A gives the mathematical theory.

1.2   PROBABILITY THEORY: A QUICK REVIEW

So much for our first viewpoint on the question of the survival of humankind. Before we move on to a discussion of the second viewpoint, it is worth refreshing some basic ideas of probability.

If you roll a fair die, the probability of getting a 5 or a 6 is $1/6 + 1/6 = 1/3$ because these two faces comprise a third of all six possibilities. This is an example of the sum rule, which states that the probability of either outcome A or outcome B is the sum of their separate probabilities $(\text{Prob}(A) + \text{Prob}(B))$. Likewise, for three possible outcomes, Prob(A, B, OR C) = Prob(A) + Prob(B) + Prob(C). (Note that the logic relations AND, OR, NOT, XOR, and so on are usually spelled with capitals in formal logic expressions.) In general, you add the probabilities of outcomes to which you are indifferent.

If you flip a coin and roll a die, the probability of getting both six and tails is $1/6$ *times* $1/2 = 1/12$. In general, the probability of multiple *independent* outcomes is the product of their individual probabilities. Independence means that the occurrence of one event has no effect on the probability of others, just as the coin and die have no effect on one another. This is an example of the product rule. For three possible independent outcomes, A, B and C, the product rule states that Prob(A, B, AND C) = Prob(A) × Prob(B) × Prob(C).

In a nutshell, AND means multiply; OR means add. Much of probability theory follows from repeated applications of these two rules. For example, let us calculate the probability of rolling a pair of dice and getting six. Suppose the dice are labeled J

The Trickiest Probability Puzzle

When probability depends on what the observer knows, one encounters some seemingly non-intuitive situations.

Suppose you are a contestant on a television show. The host offers you a choice of three boxes. One of them contains a valuable prize, but the other two are empty. Box 3 is painted your lucky color, so you choose it. To your surprise, the host opens Box 1 and shows you that it is empty. Then he offers to let you change your bet from Box 3 to Box 2. Should you do it?

For a moment you are suspicious. If you made the lucky choice, they can avoid payoff if they trick you to change your bet. But no, the show has a reputation to uphold, and surely some fans are keeping statistics. Besides, audiences like to see people win. You conclude that the offer is fair—part of the original game plan before you made your choice.

Should you stick with Box 3 or switch your bet to number 2? The solution appears on p. 26.

and K. Let the symbol $(4, 2)$ denote the probability of getting 4 on J AND 2 on K. Applying the product rule gives

$$(4,2) = 1/6 \times 1/6 = 1/36.$$

Likewise, any other combination has the probability $1/36$. So the probability of rolling 6 with the pair of dice is

$$(1,5) \text{ OR } (2,4) \text{ OR } (3,3) \ldots (5,1).$$

Finally, invoke the sum rule and replace each OR with a plus sign to get the probability of rolling a 6 with a pair of dice:

$$\text{Prob}(6) = (1,5) + (2,4) + (3,3) + (4,2) + (5,1) = 5/36.$$

#   #   #

We can apply these same rules to a survivability problem. Suppose that an entity is either type A, B, or C, with probabilities P_a, P_b, and P_c. The three types have survivabilities Q_a, Q_b, and Q_c respectively. If you draw one such entity from a random ensemble, then the probability that it is type B and that it survives to age T is

$$\text{Prob}(B \text{ AND } T) = P_b \times Q_b$$

according to the product rule. But suppose you do not care whether the entity is type A, B, or type C; you just want to know its survivability $Q(T)$ regardless of what type it is. In this case you apply the sum rule to all three products like the one above and find

$$Q(T) = \text{Prob}(A \text{ AND } T) + \text{Prob}(B \text{ AND } T) + \text{Prob}(C \text{ AND } T)$$

$$= (P_a \times Q_a) + (P_b \times Q_b) + (P_c \times Q_c)$$

$$= \text{weighted average of the } Qs.$$

The third line may be almost obvious, but it is worth noting anyhow: the survivability of a random entity is simply the weighted average of the survivabilities of the subsets in the original ensemble.

Suppose you draw samples from an ensemble in which the entities are all the same type, but you do not know which type. Then the decay you observe will be Q for the type you actually draw, not the average in the equation above. Nonetheless, this average is your best estimate *prior* to the drawing. In other words, actual statistics may deviate markedly from theoretical prior probability even when both are correct. One of the examples we have just discussed is like that. You first drew a pure isotope of ficticium, not knowing which isotope it is, and found that its decay matched one of the dotted curves in Figure 3, rather than the solid curve that represented the average. However, *prior to observation the solid curve was still your best estimate*. In this case the Ps are merely probabilities, but in a mixed ensemble they are actual abundances of the various types, in which case theory and statistics do agree.

All the theoretical curves, which began with Figure 3, are based on ignorance of real hazards. However, real hazards do exist whether we know them or not, and they produce the survival curves that we see, such as Figure 1. Thus we might reasonably expect sizable discrepancies between theory and statistics. Surprisingly, however, they conform quite well! How can we be so lucky? Very likely it happens because human intuition is quite skilled at sensing bias. In the real world, any predictable risks, biases, or cutoff ages would be well-known lore about the entity in question, in which case anyone making a theoretical model would have either chosen a different entity or modified the formulation to take the bias into account. Sections 1.5 and 1.6 below, especially the story of space-traveler Zyxx in 1.6, include just such modifications. Chapter 4 extends them to include modern man-made hazards to humanity.

1.3   CHANGING HAZARD RATES

Let us now move on and consider the case in which hazard rates are known to be changing, as they have been for humanity during the past 50 years. Calendar time T is no longer appropriate in Equation 1 since it does not reflect the accelerating risk. To quantify this, consider old-fashioned utility meters that tally consumption of water, gas, and electricity. Their dials turn rapidly or slowly in proportion to the rate of consumption. By analogy consider a virtual meter that tallies consumption of luck, in other words, risk exposure. To my knowledge there is no standard name for this quantity, so let us call it *cumulative risk*, or simply *cum-risk* (kewm-risk) for short, and use the symbol Z (as in hazard) for the meter reading. Like the utility meter the virtual meter's dial rotates at a rate that represents the current hazard rate.

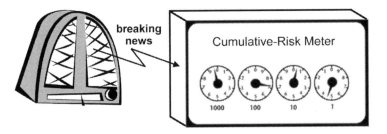

If the hazard rate is constant, our cum-risk meter turns at a constant rate; in effect it reverts to a clock, and so $Z = T$. Thus it seems likely that the generalized version of Equation 1 should simply replace T by cum-risk Z. The argument given in Appendix A.1 shows that this is indeed the case:

$$Q = \frac{J}{J+Z} = \frac{1}{1+Z/J}$$

The parameter J must also refer to cum-risk rather than calendar time.

Depending on the hazards, one can estimate Z in various ways. For example, one might keep a log of dangerous events and score each of them from 1 to 10 depending on its severity. Then Z would be the running total of all these scores. Finding a suitable Z for human survival will be the main subject of Chapter 4. That cum-risk necessarily involves world population and the world economy.

1.4 POSTERIOR PROBABILITY

As mentioned above, age is a track record for an entity's survival, but age does not appear in Equation 1 for Q. Instead, Q gives survivability from birth without reference to any later observation of age. In other words, Q is a so-called *prior* probability, which applies *before* any observation alters the odds. If we know the entity's age, we want an equation for the *posterior* probability, which applies *after* the thing is observed alive at age A. (We are stuck with these two asymmetric terms because of their long usage.)

Let G denote the posterior probability of survival, and sometimes let us expand the notation to read $G(F \mid A)$. The parentheses and the vertical bar are a standard notation from probability theory that tells us what quantities are required to evaluate G, in this case the entity's future F *after* we learn its age A. In general, $\mathrm{Prob}(X \mid Y)$ means the probability of X after we know the value of Y, in other words, the probability of X given Y.

Appendix B derives the formula for G. The trick is to calculate the formula for $Q(T)$ in two different ways and then compare the results. The first way jumps directly from time zero to T, which is Equation 1. The second way inserts an intermediate time at which an observer determines the entity's age A and inquires about its future $F = T - A$, hence

$$T = A + F$$

The result in Appendix B is the desired formula:

$$G(F \mid A) = \frac{J + A}{J + A + F} = \frac{P}{P + F} = \frac{1}{1 + F/P}; \quad P = A + J \tag{2}$$

For brevity the third expression changes $A + J$ to P for past. If a case arises where J represents gestation exclusively, then the past P is the lifetime measured from conception, unlike age A, which starts at birth. (Many papers use the letter P to denote probability, so we must be careful to avoid ambiguity. In this treatise any probability will have other letters following P.)

In the case of human survivability, J is much shorter than the uncertainty in A, too short to have an effect on the numerical estimates. Nonetheless there is good reason why we have dwelled on J in the equations above. It affects the shape of the curve in Figure 1 and in similar figures that follow. These figures in turn support the overall theory and consistency of the four approaches described in the introduction to this chapter. Leaving J unresolved would cast a shadow of doubt on our whole formulation.

Equation 2 above is related to Equation 6 in Gott's original paper [6]. Let us therefore call it *Gott's predictor*, hence the symbol G. However, he derived the equation by different means described in the next section, and his result, Equation 5, is slightly different.

A useful variant expresses the minimum future in terms of a specified level of confidence G. Solving Equation 2 algebraically for F gives the expression:

$$F = (1/G - 1) \times P$$

This form shows that an entity's future at confidence G is proportional to its past. The longer it has already lived, the longer we expect it to survive. This seems counter-intuitive only because we are accustomed to things whose vital parts wear out, or its ingredients decompose. Few of us will live 100 years, and few of our cars will exceed 200,000 miles. My prescription drug expires in three years, and milk cartons are stamped with expiration dates. Future prospects for such material objects and for living creatures dwindle with age.

Gott's predictor applies to such things *only when we have no idea what their durability might be.* However, this situation is rare because we usually have an approximate sense of the durability of physical objects: mountains last for epochs, aspirin a few years, and insects only weeks.

Prospects for other entities do indeed improve as they age. For example, ancient nations (Spain, France, China) stay intact while newer ones disintegrate: the United Arab Republic (Egypt, Syria, Yemen) lasted only three years, and both the Soviet Union and Yugoslavia were relatively short-lived. Infancy is a time of extreme danger. Many stage productions close after one performance. New businesses have high mortality, a median life of only two to four years [13, 16, 17]. (Such a short time! Do people know this when they start a business?) To reassure their customers and suppliers, old businesses advertise the year they were founded: "Serving greater Middletown since 1897." Maturity indicates that the worst hazards are under control, and their immediate future is secure.

Many entities outlast both their replaceable physical parts and the people involved. They include such things as organizations, systems, processes, phenomena, political parties, research stations, Zeitgeist, ethical standards, and extended open-ended activities like the space program. None of these entities has a characteristic lifetime: each may survive days or millennia. Hence, our theory applies well to them. It also applies to plant and animal species because their survival prospects are unrelated to the longevity of their constituent organisms. Cockroaches have produced about a billion generations and counting. By contrast, Neanderthals lasted only 10,000 generations. Like the other entities, species' future prospects improve with age as they adapt to their environment and demonstrate their ability to survive changing conditions.

Let us use 10% risk as a reasonable threshold for alarm. That means 90% survival confidence. Putting $G = 9/10$ in the equation above gives

$$F(90\%) = \left(\frac{10}{9} - \frac{9}{9}\right) \times P = \frac{P}{9} \qquad \text{That is } \ldots$$

Entity's future with 90% confidence $\geq 1/9$ of its past.

At the opposite extreme, 10% survival confidence, we put $G = 1/10$ in the equation $F = (1/G - 1) \times P$ above and find the minimum future F equals $9P$. The 90%–10% range of uncertainty from $P/9$ to $9P$ is thus a factor of 81, much broader than we would like, but the best we can expect when so little is known.

The near future in the above expression is much more important than the long term at 10% simply because the future $P/9$ is imminent, allowing little time for rescue efforts or other changes. By contrast lots of changes occur during time $9P$, which is ample to remove hazards and revise estimates of survivability.

When $F = P$, we quickly see from Equation 2 that $G = \frac{1}{2}$, which makes $P = J + A$ its median future. Accordingly, in an ensemble of entities of age A, half will survive for a future $F = P$. We shall use the following pair of benchmarks in later examples:

$$\left. \begin{array}{l} F(90\%) = P/9 \\ F(50\%) = P \end{array} \right\} \tag{3}$$

The median is one quantity that summarizes the overall size of a set of random quantities. Another summary quantity is the average, usually called the *mean* in probability theory. We tend to think of median and mean as being very similar, sort of middling. However, in our case the mean future is infinite. (Appendix C explains why, if you are interested.) This seems very odd, but a numerical simulation displays its true meaning. Pseudorandom values of G were drawn from a uniform distribution, $0 < G < 1$, and corresponding futures F were calculated using the equation $F = (1/G - 1) \times P$ above with $P = 1$. Sample sizes ranged from ten to a million. The results of this simulation appear in Table 1 below. Obviously each finite set of entities has a finite mean future, simply the sum of the futures divided by the number of them. However, as the sample size grows, the mean fluctuates randomly while increasing very slowly. There is no end to this process; the mean never converges to a finite value. In general the word *infinite* is simply an abbreviation for this sort of endless growth that never converges to a limit.

Table 1. Behavior of the mean future as sample size grows. (The number following E simply moves the decimal point. For example, $4.66E-5 = 0.0000466$.)

Sample size	10	100	1,000	10,000	100,000	1,000,000
Average future	7.40	4.25	6.35	7.03	10.66	71.38
Three least futures	0.0079 0.639 1.4697	5.63E−03 63.76E−02 4.81E−02	2.43E−03 4.16E−03 5.84E−03	6.23E−05 1.73E−04 3.06E−04	5.04E−06 2.29E−05 4.66E−05	1.88E−06 2.28E−06 3.73E−06
	⋮	⋮	⋮	⋮	⋮	⋮
Three greatest futures	8.66 21.32 32.54	35.9 44.1 88.7	414 427 799	1.00E+03 1.91E+03 7.90E+03	3.09E+04 3.59E+04 3.95E+04	1.42E+05 1.43E+05 6.01E+07

Critics have cited the infinite mean as an objection to Gott's approach. However, that is not valid as Table 1 shows. An actual statistical ensemble is always finite, as is its mean, even if the sample is a million as in the last column of the table. If there are a million humanoid species in our galaxy, there is no danger that their mean future will actually be eternal.

Moreover, Appendix C shows that the formula for the mean is just on the verge of convergence. Any slight trend toward obsolescence or any gradual decline in vitality, however small, tips the balance to a finite mean. Such aging processes are probably always present in small ways that we do not know how to formulate. Let us therefore regard mean future as a *soft infinity*. More discussion of soft infinities appears in Appendix D.2 especially the discussion of Equation D-29.

In Section 2.3 below, our database is statistics of theatrical productions in London. It includes many that are 400 years old, mostly Shakespearean. One play dates back to the 15th century. London, however, was an important town in the 10th century and surely citizens of that time performed some sort of shows on stage. Recorded drama dates back to *Play of Saint Catherine*, Dunstable, about 1110. If the mean duration were truly infinite, we should expect an occasional performance from that time, but we find none. The gap from 10th to 15th century represents a correction to the mean from infinity to a duration that is long but finite.

Likewise, few businesses are ancient. Possibly the oldest surviving corporation is the Swedish copper-mining company Stora Kopparberg (great copper mountain) which merged with a Finnish paper manufacturer in 1998 and became Stora Enso Oyj. This company was in business prior to 1288. Other ancient businesses include an Italian wine from 1385, Antinori, which has been a family business for 26 generations. If mean durations were infinite, we should expect to find surviving businesses that were founded in biblical times.

If everything loses just a little bit of vitality with age, then our theory represents an ideal age limit that is never quite attained. Real entities expire a bit sooner. To the extent that this applies to humanity, my survival estimates are somewhat optimistic.

#   #   #

This whole treatment of posterior survivability applies equally well to the case of changing hazard rates expressed in terms of cum-risk Z. Just as time T equals past plus future, $P + F$, cum-risk Z equals $Z_p + Z_f$. Then, just as Equation 1 led to 9, the equation $Q = J/(J + Z)$ in Section 1.3 leads to

$$G(Z_f \mid Z_p) = \frac{Z_p}{Z_p + Z_f} = \frac{1}{1 + Z_f/Z_p} \tag{4}$$

In other words, cum-risk Z replaces time everywhere it appears in Equation2. This is a straightforward generalization since time is simply a special case of cum-risk in which the hazard rate is constant.

1.5 PRINCIPLE OF INDIFFERENCE

Let us now proceed to our third approach to survivability questions.

A guy by the name of Guy was crossing a street one day, when he found a die lying in the gutter. Its cubic shape looked accurate enough, and its corners were not chipped, so he kept it. (It so happens that Guy was writing a book with six chapters, and the die would be a big help in organizing his material.) He saw no visible defect, and thus assumed that successive rolls would bring up each of its six faces with equal frequency. This is a classic example of the *principle of indifference*: If a process has N possible outcomes, you may be justified in assigning probability $1/N$ to each of them, especially if you look for bias that would favor one outcome over another, but find none.

Some scholars have declared unequivocally that the principle of indifference is discredited. What most of them probably mean is that nobody has mathematically defined a search for bias that fails to find any. Many of these same scholars could be caught off guard during their leisure time and enticed to play a game for modest stakes using a die of unknown provenance.

Some theorists define probability strictly in terms of the frequency of outcomes following repeated trials. For them *indifference* has no meaning. Most of us use a more relaxed definition, which allows best estimates prior to any trials. (Hopefully, we can discuss probability of human extinction without a requirement for repeated trials!)

As Guy played with his die without finding any bias, he gained more confidence in its fairness. He conjectured that it came from a discarded child's game and fell from a trashcan. Since manufacturing defects (perhaps an off-center bubble inside) are rare, Guy's confidence was high at the outset and higher after visual inspection. Yet the die might still be biased somehow. If he wanted more confidence, he could drop it in a glass of water many times to see if one number comes up too often.

The amount of confidence we demand depends on the stakes. For example, when Guy has lunch with five friends, one random member of the group chooses the restaurant. For that decision he would be willing to trust his die with no test at all. For serious gambling he would probably test the die in a glass of water. However, if diplomats are using dice to settle international disputes, they would surely use the water glass plus other safeguards against hi-tech deception. (After each session they would smash the die to show that no mechanism was hidden inside. There could be a micro-motor that moves a tiny weight off center and back. Pips on the die would double as electric contacts. A deceptive negotiator might have a dry cell up his sleeve, with which he secretly activates the motor.)

Perhaps a logician will someday take these ideas and organize a formal treatment of indifference based on confidence that approaches 100% as one test after another fails to find any bias. For now let us simply state the first of two principles to be used below in our third approach to the survivability formula:

Solution to the Trickiest Puzzle

Most people think the prize is just as likely to be in Box 2 as in 3, in which case it makes no difference whether you switch your bet—the *principle of indifference* again. We have stressed cases in which this principle holds; however, this simplistic version fails. If you switch your bet from Box 3 to 2, you *double* your chance of winning!

If you stick with Box 3, then no matter where the prize is, the host has at least one empty box he can open, which is all he needs. Consequently, his opening Box 1 gives you no information about Box 3, and so your chance of winning remains 1/3, which leaves 2/3 for Box 2. If this explanation seems too simplistic, read the detailed analysis below. In the end it uses statistical indifference to obtain the correct answer, but this time we choose the alternatives carefully.

If your initial choice, Box 3, happens to have the prize, then the host must choose which of the other two boxes to open. Let us say that he decided prior to the show by flipping a coin, which removes any chance of bias. You do not know the outcome, so from your viewpoint there are three random variables: the result of the coin flip; which box the host opens; and the important one, which box holds the prize. These have 18 combinations represented by cells in the three-dimensional array below, one table for each outcome of the coin flip. At the outset we eliminate 6 combinations, the ones labeled X0, in which the host would open the box with the prize. This leaves 12 equally probable combinations, half of which will be eliminated in the steps that follow.

Heads

Box with prize ...

Box host opens ...	1	2	3
1	X0	1/6	1/6
2	X2	X0	X1
3	X2	1/6	X0

Tails

Box with prize ...

Box host opens ...	1	2	3
1	X0	1/6	X1
2	X2	X0	1/6
3	X2	1/6	X0
Prize prob'ty	0	2/3	1/3

When you choose Box 3, that invokes the coin flip, which eliminates the two cells labeled X1. When the host opens Box 1, that eliminates the four remaining cells labeled X2 in the Box 1 columns. Six equally likely combinations remain, which are labeled with their probability 1/6. Total the columns to find the prize probabilities at the bottom right.

Note that this argument ends by applying the *principle of indifference* to the final six possibilities, but this is a sophisticated application of the principle unlike the simplistic one at the outset. This process of repairing indifference when it fails is essential to our formula for human survivability. Another example is a story in Section 1.6, the second one, in which space-traveler Zyxx repeatedly finds flaws in her assumptions of indifference, but each time she finds a way to restore it.

To finish the story, you make the mathematically correct choice and change your bet to Box 2. But you lose; by dumb luck the prize is in 3. You should have been faithful to your lucky color!

Logic alone does not guarantee the principle of indifference. However, the human mind is skilled at sensing biases (if any exist), and diligent effort to find them can make the principle workable.

#   #   #

The *principle of indifference* got its name from the famous economist John Maynard Keynes. This name suggests that somebody does not care about the outcome. This is unfortunate when applied to human survival, since most of us have a definite preference whether or not humankind survives. Of course Keynes was using the term *indifference* in its statistical sense rather than an emotional one. In prior centuries the name was *principle of insufficient reason*, coined by Pierre Simon Laplace. This name suggests that an investigation has occurred, and any reasons for possible bias were deemed insufficient. Only then are equal probabilities a reasonable default assumption. This is exactly how the concept is used here; hence the old name seems more appropriate.

Let us return to Guy when he met five friends for lunch, and he volunteered his untested die to decide which of the six would choose the restaurant. Before the roll, one of the friends lines up the others and assigns them numbers from one to six. Now suppose that the die is loaded in the extreme and usually brings up four. Is the choice of winner still fair? Yes! The unsuspecting group lined up in random order where each one was equally likely to be number four. A physical bias is not enough to nullify indifference. In this case a dishonest organizer would have to roll the die furtively a few times prior to the lineup and then take the fourth position himself. This could be very awkward if numbers 3 and 5 are having a discussion. The point is that

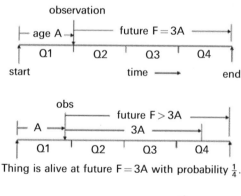

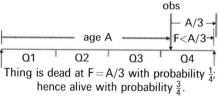

Figure 4. Timelines for Gott's survival predictor.

indifference is *robust*. More than one factor must conspire to nullify it, and this may be one reason why indifference works better than you might expect in the examples in Chapter 2.

# # #

A statistician named Stacy was strolling through an unfamiliar part of town one hot day, when she stopped at Murphy's Tavern for a pint of ale. Since she was working on survivability theory, Stacy naturally wondered about the tavern's long-term prospects. No sign in the window gave any hint. (Neither "Grand opening at our third location", nor the desperation sign "Coming soon, karaoke every Saturday night".) Stacy saw no competitor nearby, but neither did she explore all the streets in the proximity. The barmaid was a new hire with no helpful information.

On the wall inside was a photograph of the staff on their first day of business. The photo was dated, so Stacy was able to calculate the tavern's age. She has now forgotten the actual age, but no matter, she just calls it A, which becomes a parameter in her equations that follow. This was Stacy's only clue to survivability, not much, but age is a track record for survival. If Murphy's were thirty years old, Stacy would be surprised to find it expired next week. On the other hand, new businesses have high mortality. If Murphy's were only a week old, Stacy would be surprised to find it open for business thirty years hence.

While sipping her pint and pondering survival, Stacy took a napkin and drew a timeline with Murphy's opening night at the left end and its eventual demise at the right as shown in Figure 4a. Then she divided the timeline into quarters. Suppose that by some remote chance she arrived exactly at the end of the first quarter, the point labeled *observation* in Figure 4a. Then Murphy's future would be $F = 3A$, the remaining three-quarters. But the chance of that is infinitesimal because time is a continuous variable, thus never exact. To get a finite probability, suppose Stacy arrived anytime during its first quarter as in Figure 4b. The labels A, $3A$ and F would then shift in ways that make $F > 3A$. In other words, Murphy's would still be open for business at $F = 3A$.

So what is the probability of this outcome? Following Gott [6], let us invoke indifference and assign equal probabilities of $1/4$ to Stacy's arrival in each of the four quarters. This gives us a definite prediction: with 25% confidence Murphy's will be in business at future time $F = 3A$.

This reasoning aroused controversy. Steven Goodman, a professor of oncology and a biostatistician, sternly criticized it in a letter to the editor of *Nature* [18]: "If we are completely uncertain about the future [time] T, then we are equally uncertain about the cube of that duration T^3." To pursue his objection, let us define $U = T^3$. Goodman then proceeds to assign equal probability to equal intervals of U obtaining entirely different results than those of Gott. However, this is not valid. Suppose you watch a movie in which the frames are equally spaced in increments of U. At first the action is so frantic you cannot perceive what is happening. Later it slows to a normal pace, and finally to a boring snail's pace. Suppose you divide the movie's U-duration into quarters. Prior to watching it, you make a bet on which quarter has the car chase.

You would surely bet on the first quarter simply because it has most of the action. Another event that would most likely fall in the first quarter is the arrival of an observer inquiring how long the movie has been running. Thus, statistical indifference fails if it is based on U instead of T, simply because the world runs on T, not on U, and so the counterexample is not valid.

Goodman again: "there can be no meaningful conclusions where there is no information." But age *is* information. Suppose you have an atom of quacksilver, and you learn that it was created in a nuclear reactor a fortnight ago. This datum alone tells you that its half-life is probably a few days at least. Then suppose you learn that quacksilver has only two isotopes, Q^{166} with a half-life of six hours, and Q^{177} with a half-life of four years. Now you know with practical certainty that the atom is Q^{177} and you can calculate its future survival and other properties accordingly.

Goodman one last time from the same letter to the editor of *Nature*: "The labors of scientists to predict such things as the survival of the human species cannot be supplanted by statistical arguments." Wrong again. The statistical arguments are more reliable than the labors of scientists because our biosphere, technology, and behavior are too chaotic and complex for scientist's labors to produce any credible prediction. Again, try to imagine somebody in 1930 forecasting global warming, nuclear winter or genetic engineering.

#   #   #

Back at Murphy's tavern, Stacy had reasonably assumed probability 1/4 for her arrival in the first quarter of its lifespan, which implies $F > 3A$. This estimate was rough since her investigation of the tavern's business prospects was brief. Later, if she happened to overhear Murphy talking with his bookkeeper, she might alter this forecast, perhaps drastically. That posterior information would override indifference. However, suppose Stacy becomes a regular customer and chats with people, eavesdrops a bit, and explores the neighborhood. If nothing has any bearing on Murphy's survival, then she gains confidence in statistical indifference just as Guy gained confidence in his scavenged die after playing with it for a while. In like manner Section 2.2 below examines statistics of microcosms for humanity. There again we gradually build up confidence that statistical indifference applies to them and, by inference, to humanity.

Figure 5 shows the survivability curve for Murphy's, the confidence G plotted against the ratio of future to age, F/A. So far there is only one point on that curve; as discussed above it is $F/A = 3$, $G = 25\%$. Now let us find more points.

With probability 1/4 Stacy's arrival may occur during Murphy's last quarter, Figure 4c. Then age A exceeds the duration of the first three quarters, and Murphy's will have expired at future time $A/3$. The complementary outcome is that Murphy's will be in business at $F = A/3$ with probability 3/4. This also appears in Figure 4. Using the usual notation (x, y) for points on a graph, this one appears at $(1/3, 75\%)$. If we divide the timeline into other fractions, we get more points. In particular, Stacy's arrival is equally likely to occur in either half of the tavern's duration. The

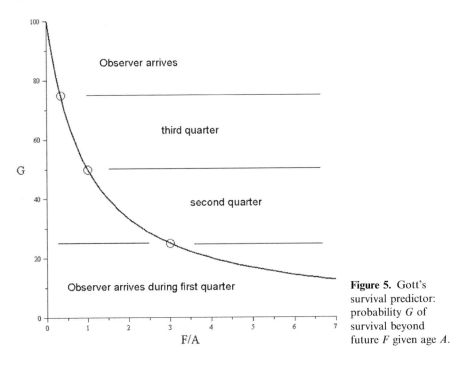

Figure 5. Gott's survival predictor: probability G of survival beyond future F given age A.

first half has probability 50%, in which case the future exceeds the second half, which lasts for time A, and thus we get a point at (1, 50%).

Obviously quarters and halves are just examples. To find a general formula, let G denote any arbitrary fraction. With probability G, Stacy arrives at Murphy's during the first fraction G of the tavern's duration, in which case its total life exceeds A/G. Subtract A, the time already lived, and the result is its minimum future F:

$$F = A/G - A$$

This is the equation plotted in Figure 5 below. In the first example above where $G = \frac{1}{4}$, the term $A/G = 4A$. If we subtract A from that then what remains is the future $F = 3A$ with probability 1/4, as before.

Solving the equation above algebraically for G yields the result

$$G = \frac{1}{1 + F/A} \tag{5}$$

This is Gott's original predictor, which is just like Equation 2 in Section 1.4, except with A instead of $P = A + J$; in other words, the gestation period J is missing. So which equation is correct? The absence of J is a paradox, but a soft one that we can resolve.

Equation 2 in Section 1.4 traces all the way back to the survival of Portuguese businesses, Equation 1. Moreover, it also fits the survival of many businesses and stage productions, as we shall see in Section 2.2 below. However, the survival of all these entities depends in part on many factors besides age: economic conditions for one, public taste for another.

By contrast, Equation 5 above is based only on hypothetical observers who inquire about age. They are profoundly ignorant of everything else, even common sense. Since they know nothing about Murphy's, they cannot know J because it is a property of Murphy's. They do not realize that Murphy's risks involve some minimum delay for hazards to develop, perhaps a month for a new competitor to open across the street, or maybe an hour for a destructive drunken brawl to develop. Nor do they realize that preparations for opening a business already comprise a track record for survival.

Since none of this went into Equation 5, we cannot expect perfect agreement with Equation 2 in Section 1.4. Clearly the latter with $P = A + J$ gives the better estimate whenever we have some idea what J might be, even a vague one. What is remarkable is that the similarity is as strong as it is, as was discussed in the last paragraphs of Section 1.2. What Gott's indifference formula has given us is another reassuring viewpoint plus the general form of the basic equation. Without Equation 5, we might not have used Equation 1 to fit the statistics. Instead we might have contrived some arbitrary fit that lacks theoretical support.

As mentioned before, gestation has no effect on human survivability because J is much less than the uncertainty in A. Nevertheless, it was worth dwelling on the subject in order to gain a sense of theoretical closure and the reassurance that goes with it.

Finally, note that the formulation in Figure 4, which leads to Equation 5, derives posterior probability (after observation) directly without reference to any prior probability. As discussed in Appendix B, the unique prior that corresponds to Equation 5 is simply $Q = 1/T$, which is improper but usable anyhow.

#   #   #

Back at Murphy's tavern Stacy stopped in the lady's room on her way out. A photograph in the hall showed what happened one Saint Patrick's Day. A mob of rowdies came wearing orange shirts. After the inevitable brawl the floor was littered with broken furniture and bottles. What would normally be Murphy's most profitable night of the year became a disaster. Since the tavern survived, Stacy estimated that the brawl should count as about four normal years of demonstrated survivability. That was just a guess based on her uncle's tavern and the troubles it endures. However, the guess is certainly better than no correction. Stacy guessed another year for gestation and incremented the tavern's age by five years in the formula calling the sum *effective survival age*.

This example suggests that Gott's survival predictor (GSP) need not be strictly limited to calendar time. The original formulation in Figure 4 was based strictly on Stacy's arrival at a random *time*, but that was before she knew anything about the tavern. Now she has a bit more information. Next let us formalize and generalize this intuitive adjustment.

#   #   #

Imagine that a race of exohumanoids has been watching Earth out of curiosity for the past two billion years. At first they stopped by every 34 million Earth years to check what geology was doing and whether the cyanobacteria (blue-green algae) had modified their swim stroke. But now with hi-tech humans racing toward the Singularity, their curiosity has peaked and they stop by every 267 days.

Whenever the word spreads that something interesting is likely to happen, knowledgeable observers come around to watch. Interesting times also tend to be hazardous. People now doing research in human survival are probably doing it only because they live in a century when our survival is threatened. Had they lived in the 19th century, chances are their thoughts would never have turned to this line of inquiry.

Recall that Z denotes cumulative risk (Section 1.3). In a stream of observers, individuals are likely to come more often when risky events are happening. In other words, observers arrive at more or less equal (on average) intervals of Z rather than T. The individual who applies Gott's indifference theory is a random member of this stream. This observer replaces Stacy's timeline for the tavern, Figure 4, by a cum-risk line, and the equations that follow remain the same except that past and future are expressed in terms of cum-risk rather than calendar time:

$$G(Z_f \mid Z_a) = \frac{1}{1 + Z_f / Z_a}$$

Here Z_a refers to age measured in units of cum-risk just as Z_p in Equation 4 refers to past $Z_p = Z_a + J$. Moreover, the equation above relates to Equation 5 just as Equation 4 does to Equation 2 in Section 1.4.

This argument based on a Z-line is rather weak. After postulating a stream of observers, it then presumes to second-guess their motives and schedule. For this aspect of our theory, the arguments in Section 1.3 above and in Appendix A are stronger. That is how it goes in this formulation: each viewpoint fills a weakness in the others. (This theory definitely is not "one for the book" in the sense that Paul Erdös used that expression [19].)

$$\# \quad \# \quad \#$$

Now we have the essential piece that was missing from the original Doomsday Argument discussed near the end of the introduction. People were applying *indifference* indiscriminately to the wrong quantities. The original random variable was our human serial number. If we divide that range into quarters like the timeline for Murphy's, then almost all of the risk falls in the last quarter. People who lived in the other quarters were incapable of self-extinction; therefore, indifference cannot apply to serial numbers. The same holds if Gott's timeline is applied to human survival: when divided into quarters, nearly all of the risk again falls in the last quarter. So again, indifference cannot apply.

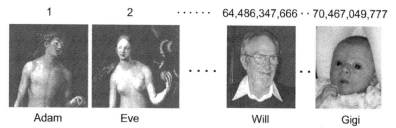

Professor Gott stated the same thing in different words. He stressed that the moment of observation must be an ordinary moment in the life of the entity in question. This excludes humankind from his formula because recent decades are far from ordinary. My contribution stems from the observation that we can revive the theory by using a measure of cum-risk instead of time. On the cum-risk scale the present moment is ordinary because equal intervals of cum-risk do entail equal risks although unequal times. Finding a suitable cum-risk for man-made hazards requires considerable research, however, and that is the main subject in Chapter 4.

#   #   #

The principle of indifference is not the only probability rule that has fuzzy conditions of validity. Another instance is Benford's law, which gives the statistics of the leading digits in numbers that are measurements of something, whether they be physical constants, scientific data or prices of merchandise. Benford's law has trapped embezzlers and tax evaders who have cooked their books with fictitious numbers that do not obey the law. Table 2 below gives the frequencies: the leading digit 1 occurs in 30% of the data, 9 in only 5%. The law has an interesting invariance: if it holds for a list of prices in dollars, then it still holds when these prices are converted to pesos or yen.

Of course, Benford's law does not hold for every entity. For example, telephone numbers fail because they do not represent measurements of anything; instead, they are assigned arbitrarily like names. There are other exceptions: heights of adult people expressed in inches, almost always start with 5, 6 or 7 simply because the range for normal adults is about 56 to 77 inches. As with the principle of indifference, you must apply some common sense before using Benford's law. In both cases, Gott and Benford, your ability to predict the law's success or failure grows as you study

Table 2. Benford's law for frequency of leading digits.

Leading digit	1	2	3	4	5	6	7	8	9
Frequency, %	30	18	12	10	8	7	6	5	5

examples and come to know bias when you see it. (In his opinion on a pornography case, justice Potter Steward of the U.S. Supreme Court famously wrote that "hard-core pornography" is hard to define, but "I know it when I see it.")

# # #

The principle of indifference, aka *insufficient reason*, may apply in different circumstances. First, the observer may be a random member in a stream of observers as discussed above. Second, the observer may have tried but failed to find clues to survival, in which case indifference is justified. This was Stacy's case at Murphy's tavern before she learned about the brawl on Saint Patrick's Day. Third, the observer may be so overwhelmed with detailed information about the entity that risk analysis becomes impossible, and again indifference is justified. Human survivability fits this last case. In all these cases one has *insufficient* reason to assign a higher probability to one interval than another.

Finally, one may *know* that a statistical ensemble of things decays according to some different formula. However, if a parameter in the formula is completely unknown, for example the half-life of a radioisotope, indifference (in the form of Gott's predictor) may still give the best prediction—temporarily. As soon as a couple of specimens expire, one can deduce an approximate parameter for the true decay formula and later revise it as more specimens expire.

The principle of indifference does not apply when you know the limits of durability. If you observe an 80-year-old man, you can be confident that he is in his last quarter of life. Hence, you cannot use the timeline argument in Figure 4 to estimate his survival because you know too much about his species. However, suppose this same man is the first earthling that a pair of visiting exohumanoids interview. If they inquire about the man's age, and if they know nothing else about the life expectancy of earthlings and how their physical appearance changes with age, then for them, statistical indifference is perfectly reasonable. The aliens' statistical ensemble is entirely different from ours. It consists of first meetings with many species scattered about the galaxy. When the old man expires at 87, the aliens get a perfectly normal datum for their research into Equation 2 in Section 1.4.

1.6 CUMULATIVE RISK

Spaceman Jorj is the leader of a colony stranded on Planet Qwimp, where the need for hydrocarbon fuel is crucial. His grandparents' generation barely survived. When his father was a child, they struck oil but not much. Now they have some industry, and life is better, but for how long? Jorj must decide how to allocate limited resources: should they develop this planet, or gamble on escape by repairing their crippled spaceship? He has no idea how much oil is ultimately available, and the colony lacks the means for large-scale exploration. But they have kept an exact tally of oil extracted; call it O_x.

Jorj recalls a drawing in his middle-school textbook, essentially Figure 4. He redraws it replacing the timeline by a line denoting oil consumption. The analogy is exact: you cannot travel backward in time, nor can you recoup consumed oil. The principle of insufficient reason applies with respect to O in exactly the manner that we applied it to time while estimating the future of Murphy's tavern. Assume that the question of ultimate oil supply is equally likely to be asked during any quarter of the oil's original volume. Hence, Jorj's best predictor for oil remaining in the ground, O_g, is GSP using barrels already extracted, O_x, instead of calendar age as the measure of cumulative risk:

$$G = \frac{1}{1 + O_g/O_x} \quad \text{or} \quad O_g = O_x \times \left(\frac{1}{G} - 1\right)$$

The second form justifies the strategy that Jorj chooses: What is available (with confidence G) is proportional to what has been extracted. He tells his people to build a maximum-security storage facility and extract as much oil as possible O_x before he commits to a decision. As soon as Jorj settles on a decision time with a value of O_x, and he decides on a confidence level G, he then calculates O_g, and converts it to calendar time using projected rates of oil consumption.

In effect the oil-extraction meter is a *luck gauge*. Jorj can think of past consumption as the amount of luck already expended. He doesn't know the total amount of luck in their future, but GSP gives him estimates of future luck based on the past. Jorj and his people may be risk-aversive and use it frugally, or daredevils and consume it extravagantly. A general discussion of time-dependent risks appears in Appendix A.1.

#   #   #

Jorj had no trouble choosing O as the measure of risk to which the principle of indifference applies. However, the choice is not always obvious, as space-traveler Zyxx learns. She parks her spaceship on Earth, hangs her universal language translator around her neck, and wanders into a nearby shooting gallery to watch earthlings amuse themselves. A patient employee explains that they have just started a contest with a valuable prize for hitting their most evasive target. In Zyxx's culture the favorite pastime is betting on the survival of things, so she gets quite involved making estimates of the target's survivability.

At first Zyxx reasons that a typical marksman gradually learns the target's evasion strategy. She expects that the target's survivability would decrease to zero more quickly than the exponential rate in Figure 2. That rate represents a constant hazard rate, but as a marksman learns the tricks, the target's risk increases. But soon Zyxx learns that targets typically survive many sessions with marksmen of differing skill and sobriety. The target has many computerized evasion programs, which management changes at random. Some marksmen are seriously trying to win. Others are taking a couple of shots just for fun. In such chaos Zyxx decides that GSP with zero gestation time is the ideal predictor, our Equation 5.

The next day Zyxx returns only to find the gallery closed. She quickly learns the weekly cycle in the earthling's calendar and discovers that the gallery operates Thursday through Sunday, but closes Monday through Wednesday, which explains the establishment's name, Four-Day Shooting Gallery. Since the target's risk alternates between zero and maximum, she cannot be indifferent in comparing intervals of calendar time as Figure 4 requires. So Zyxx uses a clock program in her pocket computer and sets it to stop when the gallery closes and to run when it opens again, thereby measuring cumulative hours of operation. Zyxx reasons that the indifference principle will apply to this measure, which she then uses to evaluate A and F in Equation 5.

Zyxx next learns that there are few contestants on Thursday morning, but many on Sunday afternoon, hence risk still fluctuates with respect to this new time although not as badly as calendar time. She learns that the turnstile at the entrance to the gallery tallies paid attendance, which she calls Y. Apparently statistical indifference applies more accurately to Y than it does to any measure of time. Zyxx revises GSP accordingly:

$$G = \frac{1}{1 + Y_f/Y_p}$$

where subscripts f and p refer to future and past. To predict the calendar time of the target's demise, she converts Y_f to approximate calendar time by using statistics of past paid attendance, which the friendly manager provides.

Zyxx is satisfied with her new cum-risk gauge (the turnstile) until she notices that some of the customers are there to attend classes for beginners, and few of them attempt the grand prize. Hence, Y is also an uneven measure of risk depending on the schedule for beginners' classes. Finally, the manager invites Zyxx to the office and shows her a counter that tallies the number of shots fired at the prize target. Clearly shot count V is the ultimate impartial gauge of cum-risk. Thus Zyxx makes her final revision:

$$G = \frac{1}{1 + V_f/V_p}$$

Again, she predicts the time of demise by converting V_f to calendar time using past statistics of the shot count, which the tolerant manager also supplies.

Zyxx has taught us that we may be justified in using indifference with respect to time until we learn that time has a bias. But then we can switch to some other independent variable carefully chosen to restore indifference. This is just the sort of reasoning we must develop to add man-made hazards to the equation for human survivability. This is the second principle on which this approach is based. In summary:

> *When bias nullifies the principle of statistical indifference, look for a different variable that spreads the risk evenly and thereby restores the principle.*

#   #   #

Clearly there is no end to the variety of cum-risks for other situations and the gauges that measure them. They are all clock-like in the sense that readings increase monotonically, but unlike a clock, their rates are not constant. Past cum-risk is the amount of luck already expended, which *improves* survivability since it is a track record for success and hence increases G through the ratio Z_f/Z_p in Equation 4, $G = 1/(1 + Z_f/Z_p)$, at the end of Section 1.4. We do not know the total amount of luck in an entity's future, but Gott's survival predictor gives rough estimates based on past consumption.

Although the ratio future/past in GSP is its only *numerical input*, GSP contains much *additional information* implied by the observer's choice of cum-risk. Over many years observers distill facts that they cannot recall in detail, but they retain common-sense knowledge that guides them to designate the appropriate quantity as cum-risk. In the case of oil reserves you know instantly to use the total volume extracted. For the shooting gallery you know after a moment's thought that shot count is most appropriate. If the gallery doesn't keep a count, then paid attendance is a reasonable proxy—but not time. For stage productions, you might do a bit of research before you realize that calendar time or performance count is best for lack of any one dominant hazard with its own characteristic timing.

Thus GSP is a highly intuitive concept well suited to the way the human mind works: it exploits the generalities we remember and the common sense we develop without demanding the myriad forgotten details we would need to set up a risk analysis. It also supplies an equation that quantifies our intuition, thus enabling estimates we could not otherwise justify.

Statisticians have long recognized and debated the subjectivity of estimation [20]. Prior to 1939, the geophysicist, astronomer and statistician Sir Harold Jeffreys [21] wrote, "The mind retains great numbers of vague memories and inferences based on data that have themselves been forgotten, and it is impossible to bring them into a formal theory because they are not sufficiently clearly stated." This vagueness did not stop him from developing useful techniques in probability theory.

#   #   #

This concludes the third approach to our survival formula, which invokes indifference. Sections 2.2 and 2.3 below substantiate the formula further with many sets of survival statistics for various business firms and stage productions. As we shall see, for every entity except one, the first 80% to expire conform to the indifference rule quite well. However, in some cases the last 20% more or less expire somewhat faster than the indifference rule predicts. Evidently small, unidentified, aging processes cause deviations from perfect indifference, which incidentally make the mean lifetime finite.

2

Confirmation

When you can measure what you are speaking about, and
express it in numbers, you know something about it; but
when you cannot measure it, when you cannot express it in
numbers, your knowledge is of a meager and unsatisfactory kind;
it may be the beginning of knowledge, but you have scarcely, in
your thoughts, advanced to the stage of a science, whatever the
matter may be.

—Lord Kelvin

Figure 6 shows a logic diagram that traces our progress so far and what to expect in Chapter 2. A continuation of this diagram, Figure 18 in Chapter 3, will complete the logic for predicting human survival.

So far our emphasis has been on Theories 1 and 2 indicated on the left; read from the bottom up. Both theories yield essentially the same formula for survival. Section 2.1 below briefly discusses Theory 3. Then Sections 2.2 and 2.3 collect the substantiating evidence indicated at the top of the diagram. The generalization from age to risk exposure has already been covered in Chapter 1 (Sections 1.3 through 1.6).

2.1 BAYES' THEORY

Our survival formula involves no modern concept. Thus it is amazing that it first arrived in 1993, tardy by at least two centuries! In the 17th and 18th centuries Blaise Pascal, Thomas Bayes, Pierre Simon Laplace, and others developed sophisticated probability theory.

If asked to develop a survival formula from frst principles, a typical statistician would think of the Bayesian approach as the most conventional way and would

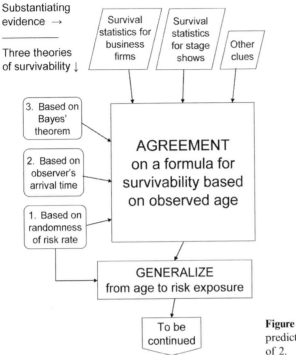

Figure 6. Logic diagram for predicting human survival, part 1 of 2.

proceed to use it. Consequently, it would seem to be the proper way to introduce the subject in Chapter 1. However, Bayesian methods have their own logic problems, and concepts are more abstruse than the ones in Chapter 1. Hence, rather than saying that Bayes confirms GSP, it seems more logical to say that GSP substantiates Bayes in our limited class of problems that concern the survival of something.

The following discussion is mostly qualitative. For mathematical details refer to Appendix D.

Consider a different probability problem. Suppose you encounter a process of fixed known duration T, perhaps a sports event or a computer program with a fixed run time. At the time of observation the process gives no hint of its start time. So you inquire, What is the probability that its past progress (age) at observation is less than some time A? Figure 7 shows the timeline. If your arrival time is random and distributed uniformly, then the probability H is simply the ratio of lengths in the timeline:

$$H(A \mid T) = A/T$$

Now express T as the sum of past and future, $T = A + F$, and then the probability becomes:

$$H = \frac{A}{A + F} = \frac{1}{1 + F/A}$$

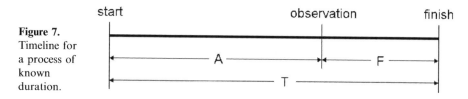

Figure 7. Timeline for a process of known duration.

which is the same as Gott's predictor (Equation 5).

We mentioned in the previous chapter that P. Buch wrote a letter to the editor of *Nature* disapproving of Gott's paper. The disapproval was motivated in part by this coincidence [14]. It looked as though Gott had solved the trivial problem above (Given T, what is the probability of A?), and then waved a magic wand to extend it to the difficult problem of prediction (Given A, what is the probability of T, hence future F?). In effect, it looks like the old sales gambit, bait and switch. However, in this case the coincidence is legitimate; no law says that two problems cannot have the same solution. Appendix D gives an extensive discussion and actually derives Gott's predictor, Equation 5, from $H(A \mid T)$ above using Bayesian theory.

2.2 STATISTICS OF BUSINESS FIRMS

Chapter 1 introduced survival statistics for expired businesses using the Portuguese data in Figure 1. This section examines more sets of data to show that they all fit Equation 1 for some value of J (except that the last survivors often die off a bit too fast as discussed below). Recall that this formula evaluates survivability at the entity's birth. This is the prior probability before it is updated by an observation at a particular age.

#   #   #

Published survival statistics are surprisingly scarce for businesses, possibly because the fates of firms are complicated: mergers, acquisitions, spinoffs, moves, foreign divisions, name changes, new owners, new management, and messy combinations thereof. Hence, the statistician must carefully adhere to some rule that defines exactly what entity she is analyzing.

A report by Baldwin *et al.* [16] is devoted to survival statistics. The other business data given here are spinoffs from other topics. Baldwin's group studied the fates of new firms in Canada. Their Table 3 reports survival of goods-producing industries, and their Table 4 reports service providers. Both tables disaggregate the data further into specialties. Figure 8 shows the aggregate of all service industries plus the two categories that deviated most from the average, "Wholesale Trade" and "Other Services".

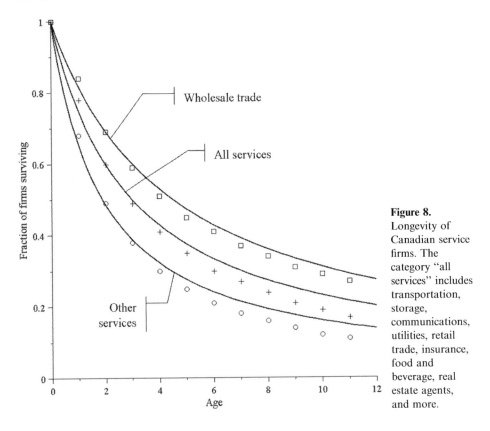

Figure 8. Longevity of Canadian service firms. The category "all services" includes transportation, storage, communications, utilities, retail trade, insurance, food and beverage, real estate agents, and more.

Gestation periods that yield the best fit seem quite reasonable for business firms:

$$J(\text{wholesale}) = 4.5 \text{ years}; \quad J(\text{other}) = 1.9 \text{ years}; \quad J(\text{all}) = 3.0 \text{ years}.$$

The quarter that live longest die off a bit faster than they are supposed to. Part of this decline may be an artifact of the study. These firms had many years in which to confound statisticians through moves, name changes, mergers, spinoffs, and the like. However, most of the die-off is probably real, a departure from statistical indifference owing to obsolescence or losses of vitality from unidentified causes. This shortfall ensures a finite mean duration. Without this extra mortality we would expect to find firms somewhere in the world that date back to the dawn of civilization, perhaps a few millennia BC.

Before continuing, let us display these same data a better way, as in Figure 9. Instead of age, we plot age + gestation on the horizontal axis. And instead of linear scales, where integers are equally spaced, we use logarithmic scales where powers of two (or any other number) are equally spaced. One advantage is that Equation 1 is always a straight line with slope -1 regardless of J. This makes the quality of the fit

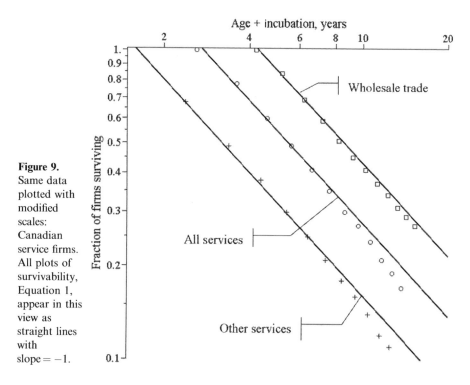

Figure 9. Same data plotted with modified scales: Canadian service firms. All plots of survivability, Equation 1, appear in this view as straight lines with slope = −1.

instantly conspicuous. Another is that the intercept on the horizontal axis is J, which makes that parameter instantly readable.

Figure 10 shows corresponding data for industries that produce goods. Again, three curves show the aggregate of all such firms plus the two categories that deviate most. The incubation periods are

$$J(\text{logging and forestry}) = 2.4 \text{ years};$$

$$J(\text{manufacturing}) = 4.3 \text{ years};$$

$$J(\text{all}) = 3.1 \text{ years}$$

One can speculate that manufacturing requires more preparation time to acquire production equipment and develop the process.

One final set of business data shown in Figure 11 compares manufacturing firms in the United States and Netherlands, the former by Dunne *et al.* [22], the latter by Audretsch *et al.* [17]. The remarkable feature is the huge contrast between gestation periods, a ratio of 3.6:

$$J(\text{U.S.}) = 2.7; \quad J(\text{Netherlands}) = 9.6$$

However, this does not affect the general behavior we see in every case: the short-lived 80% fit Equation 1 quite well, while the last survivors lose vitality.

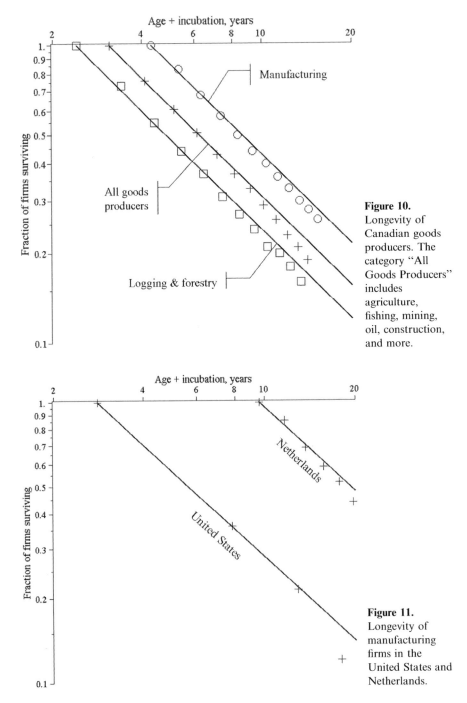

Figure 10. Longevity of Canadian goods producers. The category "All Goods Producers" includes agriculture, fishing, mining, oil, construction, and more.

Figure 11. Longevity of manufacturing firms in the United States and Netherlands.

The Dutch (Audretsch *et al.*) begin their paper with this statement: "A recent wave of studies has emerged consistently showing that the likelihood of survival tends to increase along with the age of the firm. This finding holds across different sectors, time periods, and even countries." This finding confirms Gott's theory, but Audretsch *et al.* express no hint of any fundamental significance. They treat their finding as an empirical observation possibly limited to business firms.

2.3 STATISTICS OF STAGE PRODUCTIONS

Let us now switch the paradigm to show business. Threats to a stage production are many and varied: loss of a star, critical reviews, a disaster in the theater, events that render the plot distasteful, fickle popular taste, competition, economy, or a crime wave that discourages people from going out at night. Hazards to human life are likewise varied and imponderable, and so we expect the predictor that works for the stage to work for human Gott-erdämmerung as well.

J. P. Wearing [23] has provided an excellent source of theater statistics for the London stage. His twelve volumes include every show that opened at a major theater from 1890 through 1959, if only for a single performance. Wearing's data are perfectly suited for our project, especially since he paid careful attention to small productions as well as big hits.

We must treat the data with some caution, however. For example, throughout the theater statistics there are excessive numbers of one-night stands. Evidently there are theater groups with names like *Repertory Players* that specialize in testing new concepts or in providing fillers in the theaters' schedules. Their productions are apparently not part of the main competition for attendance. It is impossible to tell how many of the one-night stands fit this category, and so we shall omit them entirely with no noticeable consequence.

There are other reasons to reject some shows from the statistical ensemble. For example, sometimes two or more short performances, rarely as many as four, are grouped together in one paid admission. For our purpose they are equivalent to multiple acts of a single production, and so I count them as only one. Occasionally, though, one of them is more popular than the others and is later revived on its own. In that case there are two entries in the statistics, one for the exceptional show and one for the remainder as a group.

Another complication is how to count shows that move to London after opening elsewhere. They acquired a survival track record at the earlier location, but not as demanding as the run in London. Wearing gives the time spent at the earlier location but not the performance count. I picked a subjective threshold for inclusion: time in London at least double the time elsewhere. Fortunately, this judgment call is not statistically significant in the final results.

Before proceeding, we must make four decisions: which years to use, what criterion will demarcate each statistical sample, what quantity will be the cum-risk, and how to define the entity in question. The chosen years sample the seven decades while avoiding World Wars, the epidemic of Spanish influenza, and the Great

Depression. The samples can be bounded either of two ways: all the shows that open in a particular period of time, which ranges from one to five years; or all the shows that were playing on specified dates. Mostly we shall use the opening date because it is easy and rather well defined, but just for comparison we include one example in which the sample consists of shows playing on specified dates.

Two possibilities for cum-risk are time duration and total number of performances. Usually they don't differ very much; most shows play about eight times per week, typically six evenings and two matinees. The better choice is performance count because that cum-risk is the greater cause of audience depletion, the number of people willing to travel to the theater and pay admission.

Figure 12 compares three choices for defining the entity in question. The entity plotted with asterisks is the *composition*, the creative work of the playwright, composer, or choreographer. The entity plotted with circles is the *production*, the series of performances produced by one team of people: managers, performers, and others working together. Finally, the third entity plotted with squares is the *run*, a series of scheduled performances with few if any changes in the theater, the cast, or management. Our theory applies to entities that have no age limit, which suggests that the best choice is the composition: Shakespeare has been playing for four centuries. Clearly it would be impossible for a production or a run to hold on that long. Nonetheless, Figure 12 serves as a sanity check to verify that the composition does conform best to our theory.

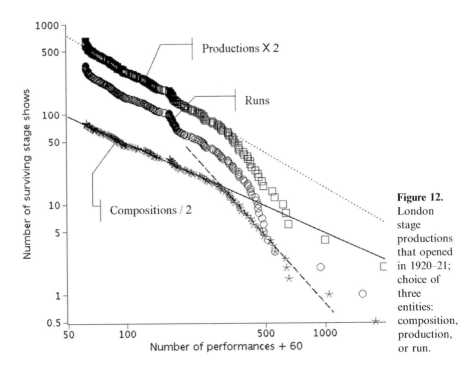

Figure 12. London stage productions that opened in 1920–21; choice of three entities: composition, production, or run.

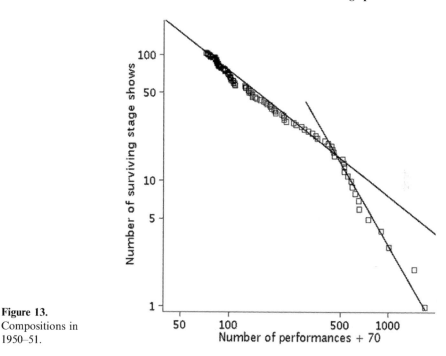

Figure 13.
Compositions in
1950–51.

Figure 12 applies to shows that opened in 1920 and 1921. The plotted survival count is doubled in the case of productions and halved for compositions in order to disentangle the curves for visibility. The solid straight line has the slope that accurately represents our theory. The compositions conform perfectly to that line until obsolescence occurs beyond 235 performances. At first glance it looks as though the change happens when about half the shows have expired, but that is an artifact of the scales. At the breakpoint, 84% of the shows have expired, so only the last 16% of them exhibit any obsolescence. As expected, the curves for productions (squares) and runs (circles) decay too fast, about the −1.3 power of the performance count indicated by the dotted line.

#   #   #

For a more recent example, Figure 13 shows survival statistics for 1950 and 1951. We would not want data from the late fifties because we might lose revivals that occurred after December 1959 when the database ended. Figure 13 shows compositions that opened in those years. The statistics are very similar to 1920–21. Beyond 398 performances, the last 15% of the shows expire faster than the indifference rate.

Anomalies do occur. Figure 14 shows what happened to shows that opened in the early 1900s. Compositions that opened in 1900 and 1901 (squares) expired too fast after only 83 performances. They comprised 42% of the total. In other words, almost half of the stage productions expired prematurely. I have no explanation except

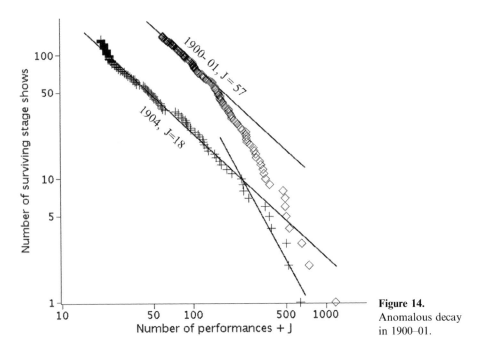

Figure 14. Anomalous decay in 1900–01.

possibly some repercussion of the Boer War. However, the decay of the first half still exhibits the slope characteristic of statistical indifference. By 1904 (crosses) the statistics had returned to normal, namely a breakpoint at 220 performances and 7% survival.

Finally, Figure 15 shows the survivability of London stage productions from 1920 through 1924. (This big ensemble, 324 shows, will be used in Section 3.3 to unmask a subtle effect that is most important to human survivability.) The main curve (squares) shows qualities very similar to the other examples, obsolescence beginning after 237 performances and affecting 18% of the stage productions.

The small curve (crosses) tests a hypothesis that the increased decay after 237 performances is not real but merely an artifact of the statistics. Perhaps we lost several long-running shows because they changed their names, moved out of central London, or were revived after 1960 when the database ends. (One of them may still be running as I write.) All or most of these losses would occur after the 1812 performances on the scale in Figure 15. Hence, the effect of L losses on the graph would be to boost the rank of number R up to number $R + L$. For this ensemble the best fit is $L = 19$, which is 6% of the total. This fraction seems unrealistically big, and besides, the warped shape of the curve (crosses) also seems unreal. It appears that the droop is not an artifact; the true survivability of these stage productions does indeed depart from statistical indifference near the end of its life.

If this is indicative of human survival, it means that the forecasts presented in Chapter 4 are overoptimistic if our species is among the longest-lived 15% of

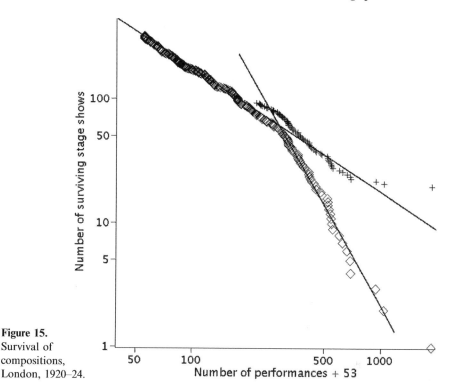

Figure 15.
Survival of
compositions,
London, 1920–24.

humanoid species. However, we have absolutely no idea what percentile we fall in until we obtain statistics for expired humanoid species (if any) throughout our galaxy.

#　　#　　#

Wearing's books include a few productions that ran for a very long time: *Peter Pan* for 3,000 performances, *Charley's Aunt* for 3,800, *When Knights Were Bold* for 2,100, and Agatha Christie's thriller *The Mousetrap* played about 16,000 times during Wearing's watch and is still open in 2009 after about 23,000 performances. For comparison, the longest recent runs in New York are *Phantom of the Opera*, 8,061 and counting (October 2007), *Cats*, 7,485, and *Les Misérables*, 6,680.

So far these super-long productions are missing from our statistics because the ensemble consists of all the shows opening in particular years, and most years do not include an opening of even one of them, much less a statistically useful sample. Thus, our data include many short-lived productions and only a few very long ones, as is apparent in Figures 12 through 15 where the data cluster into a blur at the short end, while only a few scattered points appear at the long end.

To get a better sampling, let us use a different ensemble, shows that are running on a specific date. The chance that a random observation date falls within the run time of a particular show is proportional to the show's duration. Hence, we seldom

Haymarket Theatre, London.

catch a particular (designated in advance) one-night stand, and we cannot miss a show that runs the entire decade unless the date falls on a rare night off. In this case the raw statistics do not give Q directly, but rather a different quantity from which Q can be derived. The conversion is difficult but worth doing once. It appears in Appendix E.

Let us use the year 1925 to avoid two World Wars as much as possible, although a few long runs playing in 1925 began before 1918 and a few others ran past 1939. For good luck let us use the birthday of Pierre Simon Laplace, March 23. (I would prefer Thomas Bayes, but his birth date is unknown.) A list of shows running on any date is readily available in archives of the *London Times*. Let us also use four more dates and total the data from all five samples. This gives a bigger statistical ensemble, but more important, we can equalize the intervals throughout the year to average out seasonal effects. A possible spacing is 1/5 year = 73 days, but then many of the long-running shows would fall on more than one of the five days, so for better variety let us use 4/5 year = 292 days in which case only 13% of the shows appear in more than one set. We could use wider spacing (6, 7, 8, or 9 fifths), but this would stretch out the overall time span so that more of the ensemble would occur in wartime. Four-fifths of a year seems a reasonable compromise.

Figure 16 shows the results. The initial slope is exactly that of statistical indifference (Equation 1). Afterward the decay rate increases quite abruptly to slope −2.5, a departure from indifference caused by unknown aging phenomena. At first glance it appears that about half the shows fall in the departure period. However, that is an artifact of the nonlinear scale; in fact 85% of them conform to theory, the crossover occurring at 262 performances. In this ensemble, durations extend twice as long out to 3,768 performances compared to only 1,812 during the sample of shows in Figure 15 that opened from 1920 through 1924.

I am not aware of any survival studies for business firms that select their samples as the ones open for business on a particular date. This might be a worthwhile project for readers to undertake. A good date should be remote enough that nearly all the firms have expired, but not so remote that the data are irrelevant to modern times. Perhaps 1960 would be a good choice.

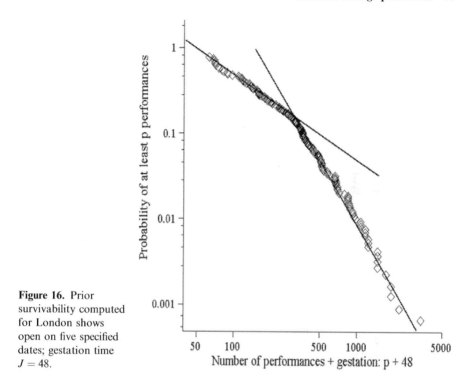

Figure 16. Prior survivability computed for London shows open on five specified dates; gestation time $J = 48$.

Wyndham's Theatre, Westminster in 1900.

2.4 LONGEVITY RANK

Consider a set of longevity statistics for N individual specimens, and let us label the longest lived with rank $R = 1$. The next longest has rank $R = 2$, and so on, progressing to the shortest, $R = N$. This concept of rank is most useful in converting statistics into probabilities and in relating durations to other statistical quantities that have been studied for more than half a century. When managing statistics on a spreadsheet, it is very convenient to have a column of ranks.

If a particular entity has rank R and duration T, then R is the number that were still alive at age T, while the statistically expected number is $N \times Q(T)$. For big N we may equate the two:

$$R = N \times Q(T) \quad \text{or} \quad Q = R/N$$

Using Equation 1 for Q tells us that

$$\text{Rank:} \quad R = \frac{N \times J}{T + J}$$

We can think of $T + J$ as duration from conception in the fuzzy sense described in Chapter 1. If we call it X, then

$$R = \frac{N \times J}{X}$$

Let us compare this to *Zipf's law* [24]:

$$\text{Zipf:} \quad R = \text{constant}/X^p$$

Here X is a random property of some object or process, and the power p is normally close to 1.0. In our case, comparison shows that $p = 1$.

For whatever it is worth, this idea of rank shows that our prior probability of survival, Equation 1, is simply the temporal version of Zipf's law.

The following are some other entities that obey Zipf's law or something quite similar [27]:

- words ranked by frequency in written language
- cities ranked by population
- businesses ranked by annual sales
- wars by number of casualties
- authors by number of books published
- bomb fragments by size
- frequency of access to web pages
- frequency of keyword usage in a search engine
- frequency of given names
- distribution of wealth

What these entities/phenomena share is scale-free behavior: the variable in question has no characteristic size or spread, in other words no mean or standard deviation.

George Kingsley Zipf (1902–1950)

Zipf was a linguist and German instructor at Harvard University. He discovered that words obey his law when ranked by the frequency of their use in written language [25]. He was independently wealthy and apparently spent his own money hiring people to count the frequency of words, a laborious task that now takes a computer a couple of milliseconds!

Many consider Zipf crazy. His book *Human Behavior and the Principle of Least Effort* [24], apparently published at his own expense, digresses in myriad wild directions including the shape of sexual organs. He notes that the annexation of Austria into Nazi Germany improved the fit of nation sizes to his mathematical law. Some think he tried to justify the annexation this way, but I think maybe he had a wry sense of humor that was too wry for some folks.

In spite of all this Zipf had moments of genius that have attracted the attention of eminent scientists. One was Murray Gell-Mann. Nobel laureate in theoretical physics, who lectured on Zipf's law at the Santa Fe insitute. Another was the renowned mathematician Benoit Mandelbrot [26], whose extension of Zipf's work is now known as the Zipf–Mandelbrot law. This is independent of the famous Mandelbrot set, the mathematician's best known work. It is a two-dimensional fractal array of points with incredible beauty. One can zoom in and see ever finer detail without ever finding a smallest element. You may view the Mandelbrot set at numerous websites, for example *http://www.youtube.com/ watch?v=gEw8xpb1aRA* and *http://en.wikipedia.org/wiki/ Mandelbrot_set*

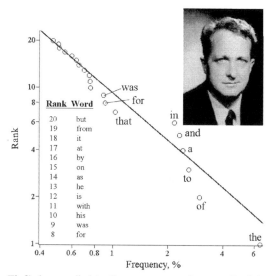

Zipf's law applied to the twenty most frequent English words from 423 short *Time* magazine articles. The equation of the straight line is rank = 9.2/frequency in %.

Time does not appear in the list above. After seven decades of Zipf, it would be amazing if nobody else has ever noticed that this law applies to durations. This is especially curious since it is easy to support Zipf when time is the variable. It is essentially Equation 1 supported by all the arguments in Section 1.1 and especially Section 1.5. However, it is not possible to extend this argument to the size of bomb fragments, annual sales, and all the other items in the list above.

2.5   UNIVARIATE SUMMARY

Several scholars have tried to refute Gott's predictor during its 16-year history. However, to my knowledge not one of them invoked real statistics like those above (Figures 1 and 9 through 17). Let us consider one last detractor, philosopher Elliott Sober [28], who published relatively recently, 2003. His paper is curious because he confesses in his Note 5 that an anonymous reviewer has essentially refuted his argument using Gott's theater statistics, and yet the journal published it anyhow.

Sober tried to make his case by noting that hazard rates of prehistoric taxa are independent of age, according to the evolutionary biologist Leigh Van Valen [29]. But this is not a counterexample. Van Valen made a hindcast, not a forecast. With data in hand he plotted survivability on a semi-logarithmic scale and obtained straight lines, the slope of which gives the age-independent hazard rate. This is exponential decay just like the thin dotted lines for radioisotopes in Figure 3. But when you make a forecast based only on the system's age, you do not yet know that hazard rate. To make an unbiased forecast, all you can do is average over the range of possible rates as shown in Section 1.2. This leads to Gott's survival predictor as shown in Section 1.1 and Appendix A.

(Hindcasts masquerading as forecasts are ubiquitous. In thousands of engineering papers the authors claim that their analysis "predicts" somebody's published experimental results. Nonsense! You cannot predict published data; you postdict or confirm them. Mere confirmation confers far less credibility than genuine prediction because the postdictor is tempted to bias his data selection or assumptions in favor of the desired result.)

Since humankind is a species, Van Valen's extinction data seem particularly relevant to our topic. As described in Appendix F, those same data intended as a counterexample in fact turn into a reassuring confirmation, but not as persuasive as our other data because a debatable assumption about sampling bias is required.

$$\#\qquad\#\qquad\#$$

This concludes our investigation of cases that have only one independent cum-risk variable, that is, univariate risk. The choice of that one variable is quite flexible. We have looked at real data for calendar time and number of performances, and at conceptual examples of shot count and turnstile count at a shooting gallery, and at extracted oil on Planet Qwimp. Next, Chapter 3 generalizes the idea to multiple cum-risks that vary independently of one another. In particular, human survivability requires two completely independent cum-risk variables: the cum-risk from natural hazards, which accumulates with time, and the cum-risk from man-made hazards, which accumulates with some other quantity based on population, technology, and economic activity. The latter has not yet been defined.

So far we have developed the univariate formula in two ways. The first appeared in Section 1.1 and began with the well-known survival law for entities that have a fixed hazard rate. Then we let this rate be completely unknown and found that the survival law transforms to our formula (Equation 1). This gives survival probability from birth, the so-called prior. Section 1.4 then used Equation 1 to find posterior probability, which applies after the entity is observed alive at a later time. For human survival that time is 200,000 years.

The second method, described in Section 1.5, dealt with the probability that the observation of age occurs at different times in the entity's life span. It gave the posterior probability directly. The two derivations agree except for small corrections that have no effect on human survivability.

Finally we substantiated the formula using survival statistics from microcosms that represent our species:

- several data sets for business firms in Portugal, Canada, and the United States
- several data sets for survival of London stage productions:
  ○ shows that opened during specified years
  ○ shows that were playing on specified dates
- survivability of prehistoric taxa deduced from the distribution of their hazard rates (described in full in Appendix F).

3

Double jeopardy

'Tis better to be roughly right than precisely wrong.
—one variation by John Maynard Keynes

For human survival we require two independent cum-risks: one for natural hazards, the other for man-made. The former is simply calendar time. Since an asteroid is just as likely to strike one year as another, its risk gauge runs at a constant rate; in other words it is a clock. For man-made risks, the virtual gauge is some imprecise measure of modern hazardous activity. It indicates serious danger due to our extreme inexperience—only a half century of coping, in contrast with 2,000 centuries of exposure to natural hazards. We must balance the two measures of risk exposure, the old one that says we are safe against the new one that warns of danger.

3.1 A PARADOX

In preparation for multiple cum-risks recall that the probability of two independent events is the product of their individual probabilities, as discussed in Section 1.2. (If you flip a coin *and* roll a die, the probability of getting both six and tails is $1/6$ *times* $1/2 = 1/12$.) In short, *and* means multiply.

Consider two kinds of entity, Jays and Kays, which have the same age and the same gestation period. Both qualify for Gott's survival predictor G, Equation 2 in Section 1.4. Suppose that Kays are exposed to all the same hazards as Jays plus an additional independent set that is equally hazardous. Since Jays obey G, then according to the product rule, the probability of Kays' surviving *both* sets of hazards would seem to be G for the first set *times* G for the second, which makes G squared (written G^2). But this cannot be because G^2 conflicts with the timeline viewpoint in Figure 4.

According to that viewpoint, Kays must obey just plain G, Equation 2 in Section 1.4, the same as Jays. How can this be, given that Kays are exposed to twice the risk?

The answer is that we cannot infer Kays' survival statistics from Jays' because on average the Kays are a hardier lot by survival of the fittest. Most frail Kays have already succumbed to the double set of hazards prior to observation. Thus G remains valid for the Kays survival even though the risks are double. To apply the product rule separately to the Kays' dual sets of hazards, the correct breakdown is the square root, $\sqrt{G}$, for each set, which then multiply to give the required G for exposure to both sets. If Jays and Kays exchange risks, then the hardy Kays are quite safe while the frail Jays are at great risk.

> **Public Health**
>
> The story of the Jays and Kays in Section 3.1 and survival of their fittest has implications for public health. Popular medical literature has mentioned that immigrants to North America from underdeveloped countries are extraordinarily hardy, and some commentators regard this as a puzzle. But the explanation is simple: Like the Kays, these immigrants are survivors of societies that had poor medical care for many generations. The long-term implication is that public health may depend not so much on medical advances, but rather on the *rate* of medical advance keeping pace with the *rate* of genetic atrophy!

3.2   FORMULATION

Suppose that a certain impresario wants the most accurate formula he can get for survival of stage productions. One big risk, probably the greatest, is the show's *popularity* for which parameter J is small, about the time it takes to write a review, have it published, and for the public to read it. The impresario lumps all other hazards into *miscellany*, which includes finance, personnel problems, casting, theater, stage, and the risk of losing the show's star. Many of these hazards begin months prior to opening night.

The formula has two gestation periods, one for each set of hazards; call them J_{pop} and J_{misc}. The impresario reckons that the two sets are statistically independent; neither has much influence on the other. In that case, as we have just discussed, the probability of surviving both is the product of the probabilities of surviving each by itself, which is the product rule discussed in Section 1.2. Thus the prior probability of survival from opening night disaggregates into two factors:

$$Q = \left(\frac{J_{\text{pop}}}{J_{\text{pop}} + T}\right)^{1-q} \times \left(\frac{J_{\text{misc}}}{J_{\text{misc}} + T}\right)^{q}$$

By calling the exponents q and $1 - q$, we ensure that their sum is 1.0 as required, Section 3.1 In another application of this formula, the two gestations might be equal, $J_{\text{pop}} = J_{\text{misc}} = J$. Then the two factors coalesce, Equation 1 is restored, and nothing has changed.

The two exponents express the relative importance of their respective hazards, and so our impresario needs some means to evaluate q. He makes a list of expired shows and uses his insider's knowledge of showbiz to tabulate the cause of each demise. Then he tallies the number of shows that fell victim to each set of hazards.

One might expect that exponents q and $1-q$ are in the same ratio as the respective tallies of expired shows, and indeed they are for the special case of equal gestation periods. However, in general the hazard with less gestation has extra mortality since the hazard acts early when its victims are more vulnerable. Appendix G works out the full theory.

After a stage production survives to age A, its posterior survivability for future F is a revised GSP derived in the same manner that Equation 1 led to Equation 2. The result is

$$G(F\,|\,A) = \left(\cfrac{1}{1+\cfrac{F}{J_{\text{pop}}+A}}\right)^{1-q} \times \left(\cfrac{1}{1+\cfrac{F}{J_{\text{misc}}+A}}\right)^{q}$$

Both factors in the impresario's predictor involve measures of calendar time, but this is not a restriction. In our later applications one of the cum-risks will be time, and the other(s) something quite different, perhaps one of those in Section 1.6. Let us call this cum-risk Z (for hazard) as before. For example, Z might run in spurts like the shot count at the Four-Day Shooting Gallery. Although the statement above, "this is not a restriction", seems intuitively obvious, it lacks the rigor of a mathematical proof. Therefore, we shall spend considerable effort to corroborate this statement with statistical data, and this appears in Section 3.3.

For now, assume that this generalization holds. Then the modified version of prior survivability, Equation 1, becomes

$$Q = \left(\frac{J}{J+T}\right)^{1-q} \times \left(\frac{K}{K+Z}\right)^{q} \tag{6}$$

and the posterior survivability or GSP follows as before, Equation 5, except this time there are two futures, T_f and Z_f, and two ages, T_a and Z_a.

$$G(T_f, Z_f\,|\,T_a, Z_a) = \left(\cfrac{1}{1+\cfrac{T_f}{J+T_a}}\right)^{1-q} \times \left(\cfrac{1}{1+\cfrac{Z_f}{K+Z_a}}\right)^{q} \tag{7}$$

Equations 6 and 7 resemble the impresario's equations, the two above Equation 6, as you would expect.

#   #   #

Most hazards that we deal with are statistically independent. For example, the chance that an asteroid exterminates humanity is unrelated to the chance that a mad scientist does it. There are exceptions. For example, the chance that nature kills humanity with a new contagion is related to the chance that mankind provides the rapid transit systems that defeat quarantine efforts. On the other hand, it is also related to the chance that man develops a vaccine in time. In the first exception the two effects cooperate; in the second they oppose. We have no reason to think that positive correlation is more or less likely than negative. Hence, in accord with statistical indifference, let us assume that hazards are statistically independent on average. Then the product rule applies again as it did for the impresario.

We can expand these ideas and factor GSP into a product of probabilities of many independent risks or categories of risk. In principle there could be dozens:

$$G = G_1^\alpha \times G_2^\beta \times \cdots \times G_{24}^\omega = \left(\frac{1}{1 + F_1/P_1}\right)^\alpha \times \left(\frac{1}{1 + F_2/P_2}\right)^\beta \times \cdots \times \left(\frac{1}{1 + F_{24}/P_{24}}\right)^\omega$$

(8)

where the exponents sum to 1.0 ($\alpha + \beta + \gamma + \cdots \omega = 1$) and the pasts P include any applicable gestation periods.

In practice, however, the limited accuracy of input data would seldom justify more than two factors or maybe three at the most. Any further disaggregation would be counterproductive because one cannot establish the many parameters (exponents and gestations) accurately enough to realize full theoretical precision.

We can understand the rather formidable-looking Equation 8 from another viewpoint. Each parenthesis without its exponent is a GSP for exposure to one hazard. The value of each GSP derives from a risk gauge that disagrees with all the others. The only impartial resolution is an average over the set of all such GSPs. However, this must be a weighted average that favors one hazard over another in proportion to its severity, namely $\alpha, \beta, \ldots \omega$. Equation 8 is just such an average, a geometric one. A different type of average (for example, arithmetic or harmonic) would be inappropriate owing to the multiplication rule for the probabilities of independent hazards.

\#   \#   \#

Equations 6 and 7 apply to humanity's survival. The first factor represents natural hazards. Since a bolide (meteoric fireball) is as likely to hit one year as another, the cum-risk is

---

Algebra Review: Fractional Powers

In preparation for this section you may want to review fractional powers and their products. The square root is the 1/2 power:

$$G^{0.5} = \sqrt{G}; \quad \text{hence,} \quad (G^{0.5})^2 = G$$

This is a special case of the rule that powers of powers multiply:

$$(G^r)^q = G^{r \times q}$$

The zero power of any number is one:

$$G^0 = 1.0; \quad \text{hence,} \quad G^0 \times Z = 1.0 \times Z = Z$$

The first power of any number is just that number itself:

$$G^1 = G$$

When different powers of the same number multiply, their exponents (superscripts) add:

$$G^r \times G^q = G^{r+q}$$

In our case r and q represent the relative severity of two sets of hazards. For the Kays in Section 3.1, r and q are both 0.5 so that

$$G^{r+q} = G^1 = G$$

When the two severities are unequal, their sum must still be 1.0.

Therefore, we replace r by $1 - q$ so that the sum 1.0 is automatic:

$$G^{1-q} \times G^q = G^1 = G$$

This subject arises again in Chapter 4, where GSP for human survival is the product of two factors: one for natural hazards based on calendar time, and the other for man-made hazards based on world population and technological progress.

The Birthday Puzzle

Probabilities are sometimes counterintuitive. Consider a group of people gathered in a room. The probability that two or more have the same birthday is 51%. How many people are in the room? Make the obvious simplifying assumptions: no twins, and the birthrate is constant throughout the year, except February 29 when the birthrate is zero.

As is often the case, the first step is to change the question and ask, what is the probability that no two people in a group of N have the same birthday? Then search for the value of N that gives 49%. Think of people entering the room one at a time. The chance of the second being different from the first is 364/365. If those two are different, two days are taken, and so the chance of the third being different from both of the first two is 363/365. The probability of both the second AND the third being different from the first AND from one another is the product of the two fractions, because AND means multiply. And so on. For N people the answer is

$$\text{Prob(no two)} = \frac{364}{365} \times \frac{363}{365} \times \cdots \times \frac{366 - N}{365}$$

Note the resemblance to Equation 8, another series of AND probabilities. It is easy and amusing to set up this series of products on a spreadsheet.

The answer is 23. That's right, in a gathering of only 23 people, the probability that two or more have the same birthday is 51%. Most folks think the number should be greater. They tend to think that random numbers are nearly uniformly distributed, which they are on average. But each independent trial has by chance many clusters, which produce the birthday coincidences. Shake a handful of coins and dump them on a table. Then look away (or take off your glasses) and arrange them in a row. You may be surprised how often three or four consecutive heads or tails occur. The gambling industry makes a good profit by exploiting just such misconceptions.

One final datum: in a room with 41 people, the probability that two or more have the same birthday has risen to 90%.

simply time. The other cum-risk represents man-made hazards with Z to be defined later. This hazard was dormant for most of human history but is now accelerating rapidly. However, the first factor gets the greater exponent because nature has many times demonstrated her ability to extinguish species on a massive scale. By contrast, humans may not be capable of self-extinction over the period for which this study is valid. In other words, q is small.

3.3   AN EXAMPLE

For the univariate case, Sections 2.2 and 2.3 show numerous examples that substantiate the basic formula. We need similar examples for the multivariate case because the formulas have a new aspect, the (usually uncertain) exponents in the equations in Section 3.2 and the claim that their sum must be 1.0. Without examples one has an uneasy feeling that understanding is incomplete. To fill this gap we need survival statistics for entities subject to dual cum-risks, one of which is not time. We prefer that the statistical records state the cause of each demise so that we can fit the formula to multiple equations, one for the "body count" due to each hazard; see Appendix G. Let us call this a *strong substantiation*.

If we were doing a medical study, we could easily obtain strong substantiation, namely mortality data disaggregated by cause of death. All we would need is a stack of death certificates. However, for our class of entities, the ones with no characteristic longevity, statistics listing causes are surprisingly scarce. Without them we can still get a *weak substantiation* by adjusting all parameters to find the best fit to one equation, the overall survivability, Equation 6 or 9. If the resulting set of parameters is both plausible and unique (only one best fit) we then have reasonable assurance that the formulation is working.

We do have one example, unfortunately the weak kind. However, the answers are plausible, unique, and surprisingly consistent using two different criteria for data selection. Again the entities are London stage productions, but in this case the risk splits into two parts, the overall duration T and the overall number of performances S (for shows). Survivability declines with S simply because the show depletes its supply of people willing to travel to the theater and pay admission. Excessive duration T also stresses a show. One with frequent hiatuses fails to provide steady employment for the cast and staff, and so startup costs recur with each revival. In the long haul, public tastes change or events redirect public interest.

Appendix H optimizes the parameters J, K, r, and q in the following equation to give the best fit to the theater data:

$$Q = \left(\frac{J}{J+T}\right)^q \times \left(\frac{K}{K+S}\right)^r \tag{9}$$

(The optimization actually includes another term in the denominators of this equation, which allows for some obsolescence in long runs, but that is a detail described in the appendix.)

Equation 9 is like 6 except that the sum of exponents is *not constrained* to be 1.0. The intuitive argument near the end of Section 3.1 says that the sum of exponents $q+r$ must equal 1.0. However, that argument does not meet rigorous standards of mathematical proof. Therefore, instead of forcing the condition that $r+q=1$ as in Equation 6, we let the optimization process discover it, thus enhancing our confidence in the formulation. For the better of two sets of data, the sum $r+q=1.013$; for the other a respectable 0.945. This is a most important result because two factors like those in Equations 6 and 7 are fundamental to the calculation of human survivability, one factor for natural hazards, the other for man-made. Now the statistics corroborate it.

#   #   #

The statistical ensemble for this example consists of shows first performed in London during the five years 1920 to 1924 (World Wars avoided again). Figure 17 shows a scatter diagram of each production's performances and durations. Most of them fall on or near the upper straight line, which denotes eight performances per week, typically six evenings and two matinees. For this majority, the two cum-risks are locked in sync. Only their incubation periods J and K may differ significantly. For lack of samples in the direction perpendicular to the lines, one might expect the statistics to discriminate poorly between them. However, the ensemble is big, 310 stage productions, and this suffices to discriminate well enough despite the synchronization. For these shows, $q+r=0.945$, which differs from the ideal 1.0 by 6% instead of 1% for those off the main sequence.

The few shows to the left of the solid line in the scatter diagram played twice per day during at least part of their run. The dashed line represents an average of two

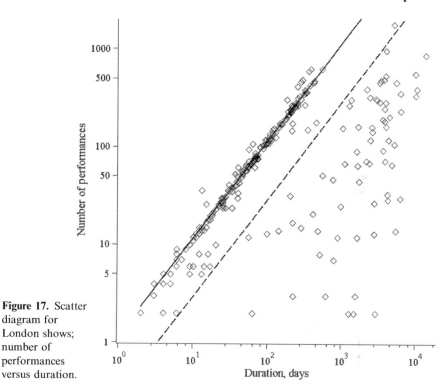

Figure 17. Scatter diagram for London shows; number of performances versus duration.

performances per week. The 69 shows to the lower right of this line (22% of the total) have lengthy hiatuses. In some cases a brief run is an annual event, sometimes during the holiday season. One might regard this group of *occasional shows* as an entity distinct from the majority in the *main sequence*. Perhaps each occasional show has a small but devoted following of people who expect to attend it repeatedly, as many opera and ballet fans do. These shows discriminate very well between the two cum-risks.

Recall that the two factors in Equation 9 derive from the product rule: if X and Y are *statistically independent*, then the probability of both equals the probability of X *times* the probability of Y. One might think that the strong correlation between performances and duration along the main sequence would violate the requirement for independence, but this is not the case. The kind of independence that the product rule requires is that the *hazards* against performances (for example audience depletion) are unrelated to the hazards against duration (for example revival startup costs). The product rule does not exclude schedule synchronization, which is a different sort of dependence.

Appendix H describes the computation of parameters J, K, q, and r for Equation 9 that best fit the statistics. Two tables below list summary results for the two ensembles, occasional shows and main sequence. Table 3 gives the standard deviation

Table 3. Standard deviations from theory, %.

	Bivariate	Univariate
Occasional shows	1.52	3.10
Main sequence	1.37	1.41

Table 4. Summary quantities.

	Shows	Perf. wt, r	Dur. wt, q	r/q	$r + q$	J, perf. median	K, dur. median
Occasional	69	0.643	0.370	1.74	1.013	159	4.1 yrs
Main sequence	310	0.864	0.081	10.7	0.945	49	3.8 yrs

from theory. One column applies to our bivariate case. As a reality check, the column on the right shows that the fit degrades if we revert to a univariate fit using only performance count, as in Section 2.3. All of these fits are well within the expected range for the sample size; see Appendix H. The main sequence is a closer fit simply because it is a bigger sample, 310 stage productions instead of 69.

Table 4 shows other summary quantities. Consider the ratio of statistical weights $r/q \approx 1.7$ for occasional shows. This says that the number of performances is a weightier survival predictor than all the temporal hazards. This is plausible because performances deplete the supply of paid admissions, the most fundamental survival issue for any stage production. The ratio 10.7 for main sequence productions indicates that duration has little effect on survival, probably because they have no recurring startup costs.

The long median for occasional shows, 4.1 years, probably indicates that they are well adapted to long durations. The sum of exponents $r + q$ confirms the theoretical value 1.0 as noted above. In summary, the results pass all the sanity checks.

3.4   LOGIC DIAGRAM

Figure 6 at the beginning of Chapter 2 showed the first part of a logic diagram for calculating human survivability. The diagram ended with a formula having a single cum-risk, the so-called *univariate* case. The second and final part of the logic diagram appears in Figure 18. The two parts are combined in Appendix I and converted to an outline format with more details added.

At the top left of Figure 18 the univariate results are generalized to *bivariate*, the process completed in Section 3.2. The top right corner of the diagram denotes the reassuring example treated in Section 3.3 above. It corroborates the formulation, especially the important claim that the sum of exponents must be 1.0.

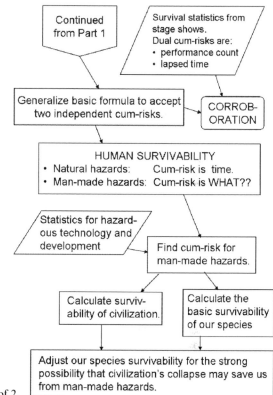

Figure 18. Logic diagram, part 2 of 2.

In the next chapter we shall move on to address directly the question of human survivability as shown near the center of the diagram. We have noted before that for natural hazards the indifference principle applies to calendar time: a bolide is as likely to strike Earth one century as it is another. Choosing a cum-risk Z for man-made hazards is our major remaining task. It must somehow involve economy, technology, and population.

Before doing that, however, let us end this chapter with a brief discussion of some other possible avenues for future research that readers may wish to take on.

3.5   FUTURE RESEARCH

Gott's survival predictor is supposed to be a fallback formula when the observer knows nothing but the entity's age. However, in the univariate case we find that real statistics from a variety of sources obey GSP quite well, Figures 1, 8 through 11, 13, 15, and 16. The question now arises: Does this bonus hold in the multivariate case? Doubt arises since the formula has a new aspect, the uncertain exponents in

Equations 6 through 9. If the individual specimens in a statistical ensemble have a variety of exponents that conform somewhat to our weighted average, then the fit will be fairly good. But if the true exponents are all about the same, and we average over different values only out of ignorance of the real value, then the fit will be poor. If possible we should learn which case arises most often in practice. The example in Section 3.3 is quite encouraging, but still, it is only one example; furthermore, it is a weak substantiation in the sense defined there. A couple of strong substantiations would improve confidence.

Consider for example the preservation of historic structures. The cause of each demise is normally available in public records. Cum-risks are time (fire, cyclone, other natural hazards), public apathy (inferred from the number of visitors recorded in the visitor log), official neglect (frequency of maintenance, security measures), and pressure to replace the structure with something new and profitable (market value of the land). To estimate the survivability of a single structure, this is a good application of the theory; see Appendix J.

However, it would be awfully labor-intensive to acquire a sizable database in order to substantiate the theory for the case of historic structures. This would require research at each site to examine such factors as visitor admissions, history of land prices, and fire protection, which vary radically from site to site. For example, replacement pressure on a historic structure in downtown Chicago vastly exceeds the pressure on an abandoned gold mine in the Mojave Desert. A minimal ensemble, say sixty historical structures that have been destroyed, could turn into many thesis projects for students of futurology or forecasting.

Other qualified entities are difficult to find. If you want to take the challenge, the following list of failed attempts will help you avoid some pitfalls:

- Business exits—Besides time, what is the second cum-risk? An outsider rarely has the full true story. This is in contrast to show business, which thrives in the proverbial fish bowl.
- Road show closings—The two cum-risks would be number of performances and number of moves, but we should move on from stage productions to something else.
- Computer hard drives—Cum-risks are hours running and number of start–stop cycles, which are almost never recorded. Besides, the drives are physical devices that wear out after a characteristic longevity.
- Shelf life of library books—Even though paper has a natural life, it may be long enough to use this example, at least for books printed on acid-free paper. Cum-risks are calendar time (obsolescence, vandalism, bookworm attack) and number of times loaned (not always returned). I obtained statistics for hundreds of copies of Harry Potter novels at the San Diego public library, but loss by the readers' failure to return books was so dominant that I could not detect any other hazard in the statistics.

4

Human survivability

> *Man is only a reed, the weakest in nature; but he is a thinking reed. There is no need for the whole universe to take up arms to crush him: a vapor, a drop of water is enough to kill him. But even if the universe were to crush him, man would still be nobler than his slayer, because he knows that he is dying and the advantage the universe has over him. The universe knows nothing of this.*
>
> —Blaise Pascal (1623–62)

We may be the only humanoid species in our galaxy, or we may be one of millions. In any case, let us pretend that a substantial number have come and gone. This lets us use their presumed existence to retain ordinary concepts of statistical ensembles.

Most humanoid species that have arisen in the past have already expired for various reasons, as have the vast majority of earthly species, often after a run of one or two million years. Some of those extinction events had natural causes, others were self-inflicted. Some of the latter extinctions happened suddenly in high-tech accidents. A few species succumbed to sabotage by crazed individuals, the proverbial mad scientist. Others were doomed when they triggered runaway processes (positive feedback) that poisoned their biosphere, changed their climate, or allowed voracious self-replicating things, either biological or robotic, to overrun their planet. Whatever the cause, their luck eventually ran out.

Imagine that we somehow obtain the *Grand Galactic Book of Knowledge*, which contains histories of every humanoid species that has ever lived and died anywhere in our Galaxy—one million in all. After the enormous excitement subsides, statisticians process the data. First, they choose a subset of 10,000 species that seem most human, lived on Earth-like planets, and are roughly our age, 200,000 Earth years (adjusted for each species' metabolic rate).

In one project, statisticians choose a preindustrial date, say 1000 AD, and look for matches to our situation in the extraterrestrial data. They find a subset of 100 species. Pretending that this time is the "present", they look at survival statistics for "future" time F. In other words, the statisticians simulate an earthling living in Y1K, and then they inquire about survival using galactic statistics instead of theory. This is a means for testing Gott's original simple predictor. Agreement with galactic statistics ought to be quite good in view of the many confirmations summarized in Sections 2.2 and 2.3.

Next, these statisticians pick a different subset of 100 humanoid species, the ones whose civilizations most closely match ours in 2009. There is some overlap with the first subset, perhaps a third. In the randomness of events most species that seemed similar in Y1K eventually drifted away from our situation, while others that seemed different drifted in.

Again the statisticians run the numbers and compare them to the generalized GSP developed in this section. This time they do not agree as well. Compared to the preindustrial case, two difficulties have arisen. One is assignment of statistical weights to natural hazards as compared to self-extinction, the quantity q in Equation 6. To understand the other difficulty, recall the example of the Four-Day Shooting Gallery in Section 1.6. Our observer Zyxx went through three flawed cum-risks before he finally found the optimum quantity for statistical indifference, namely shot count. Hazards for self-extinction are more complex, and it is tricky to find the cum-risk that best represents the full set.

In summary, we had only one level of uncertainty for the preindustrial date; namely, on the galactic survivability curve, what percentile do we fall in? That was all, because we could be reasonably confident that the galactic curve was quite similar to Gott's predictor. Now, however, we still have the percentile uncertainty, plus a second level of uncertainty about the parameters within the formula.

$$\# \qquad \# \qquad \#$$

Imagine a rare combination of disasters that kills 99% of the world's population. It might be a coincidence involving volcanism, extreme weather, and mutant mosquitoes, followed by wars fought over remaining resources. This would be the worst calamity that humankind has ever suffered by any criterion—except one. For long-term species survival this is the best that could happen! In its aftermath the 66 million survivors are a hardy lot, wiser for the experience and disinclined to rash judgments and dangerous behavior. In remote areas isolated tribes are safe from epidemics, drugs, and other man-made hazards that afflict civilization. The biosphere gets relief from human stress. Carbon dioxide concentrations drop to normal. Nobody is stressed by crowding. In summary, the man-made hazards practically vanish.

As this extreme example shows, what's good for our species' survival is not necessarily good for society, civilization, or individuals, nor does it conform to accepted ethics and political rectitude. Throughout this section we must retain this detached viewpoint, focus on survival, and avoid lapses into conventional thinking about the well-being of society and individuals.

An early version of my manuscript was reviewed by Theodore Modis, a physicist turned futurologist and strategic business analyst. Dr. Modis slogged through my mathematical details and made several helpful suggestions. However, at this point he lost his focus:

> "Why should all people do nothing but damage? There are people whose net contribution to the survival of society is positive. ([Wells] could be one of them by ... raising public awareness!) One may defensibly argue that people in general do more good than bad toward the survival of society."

Whoa! The subject is not about good, bad, nor society. The subject is the probability of an extinction event. The potential causes are utterly indifferent to our social values and cultural norms. Besides, good people do not offset bad ones. Recall the diligent workers in airport security on the morning of September 11, 2001. The final calamity will probably blindside us, so good folks working for good social causes are ineffective. Also, the modest positive efforts of well-meaning citizens are offset by the fact that they too generate carbon dioxide, spread pathogens, consume nonrenewable resources, and unknowingly purchase goods from industries that pollute.

Table 5 lists some extreme examples to emphasize the disparity between people's goodness/badness and their impact on our species' survivability. The last column contains remarks purposely devoid of human compassion. (Please read the disclaimer at the top!) The correlation between acts that are good/bad for society and those that are good/bad for survival is near zero, perhaps slightly negative.

Clearly, if we are engaged in trying to calculate the probability of an extinction event we must not be distracted by conventional ideas of good and evil. We must view this whole subject with the detachment of an exohumanoid ethologist visiting Earth to do field work.

4.1 FORMULATION

Let us again compare risk measurement to the meters that public utilities install to measure consumption of gas, water, and electricity. In our case they register consumption of luck, meaning exposure to hazards. For human survivability we use two virtual meters. The hazard rate for natural hazards is fairly constant, and accordingly its meter is simply a clock that registers calendar time, the quantity in Gott's original formula. This clock now reads age $A = 200,000$ years [30]. If this number and GSP had been published in 1900 prior to our ability to commit self-extinction, our ancestors would have used Equations 3 and found that we have a 10% risk of Gott-erdämmerung in 22,000 years. Little more could be said at that time, but now things have changed drastically. Calendar time is no longer the dominant cum-risk.

Starting some time in the mid-20th century, concern shifted from natural threats to man-made hazards: first nuclear winter, then greenhouse emissions, and now

Table 5. Dispassionate look at social values. Approved humanitarian behavior is unrelated to survivability of our species. *Disclaimer: The viewpoint listed in the last column, which is devoid of human compassion, is not a suggestion for public policy!*

Person	Goodness or badness	Impact on human survivability
Charismatic leader of a suicide sect	Lunatic, killer	Good: he takes out some irrational people who may be prone to drastic behavior. They or their descendents could harm us all
Participant in Spaceguard Survey, which tracks asteroids and other Earth-threatening objects	Nice nerd	Earth's protector
Surgeon who repairs serious congenital defects in infants	Wonderful healer!	Dysgenic, instrumental in propagating defective human genes
Foreign-aid worker helping tropical people eradicate mosquitoes	Humanitarian	Negative, those mosquitoes are nature's guardians of undeveloped land and biodiversity
Joaquín Balaguer	Evil ruthless dictator of Dominican Republic	Environmentalist and savior of Dominican forests. He expelled all forest encroachers from squatters to lumberjacks to wealthy owners of mansions
Saddam Hussein	Cruel dictator	Ecological disaster: set fire to oil wells, spilled oil in the Persian Gulf, and drained swamps in southern Iraq killing species

biological warfare and genetic engineering. We must devise the second meter that accelerates with world population and advancing technology and thus registers our exposure to man-made hazards. The two come together in Equation 6 for Q, the joint probability of survival prior to any observation of age. Just as it has two factors, one for natural hazards and another for man-made, the corresponding predictor after age observation will also have two factors. If we let f stand for both futures, time and cum-risk, and let p stand for both pasts, then the formula for human survivability is simply

$$G(f \mid p) = G_n^{1-q} \times G_m^{q}$$

The expanded form of these two factors appears in Equation 7. Recall the notation in parentheses on the left: $(f \mid p)$ denotes the probability of future f if the past p is known (given). The first factor, G_n, stands for the old <u>n</u>atural risks, while G_m represents new <u>m</u>an-made, <u>m</u>an-activated and <u>m</u>an-exacerbated hazards. The exponents in this equation and in Equation 7 total 1.0, as we learned in Chapter 3 that they must. The limit $q = 0$ is the case in which man-made hazards vanish; $q = 1$ is the limit where natural hazards vanish.

The equation for G_m has the form in Equation 4:

$$G_m = \frac{Z_p}{Z_p + Z_f} = \frac{1}{1 + Z_f/Z_p} \tag{10}$$

in which Z (as in hazard) is the cum-risk, which is not yet defined.

Let us change the notation for past and future time:

$$\text{age } A \text{ or past } P \rightarrow T_p; \qquad \text{future } F \rightarrow T_f$$

where subscripts p and f denote past and future. The reason is to make the time symbols F and P in Equation 2 correspond to the Z symbols; hence,

$$G_n = \frac{1}{1 + T_f/T_p}$$

The new notation puts emphasis (capital letter) on the two cum-risks, time T and the new one Z.

#   #   #

The next task is the hard one, defining an appropriate cum-risk Z for use in Equation 10. Throughout prehistory, ancient times, and part of the twentieth century, Z barely budged from zero because humans were then incapable of self-extinction. In the twentieth century we began making increasingly dangerous stuff, and so Z took off and has accelerated ever since. When we do something risky, Z tallies our consumption of luck just as the oil meters on Planet Qwimp tally consumption of petroleum.

Consider the increment ΔZ by which Z increases in a single year. One factor in ΔZ must be world population p during that year because p is the pool of potential perpetrators, whether extinction is the work of a single mad scientist or the result of everybody's collective bad habits. So whom do we count in this population? Everybody? Or only those deemed most dangerous? Perhaps we should count only the populations of industrial nations that have the most power to commit extinction and also generate the most greenhouse gas.

Opinions differ. Those who worry about disease should count people in the most crowded germ-breeding areas of Africa and China. Those who fear robotics running amok should count the world's myriad sweatshops that make it affordable. Those concerned about deforestation should count everybody: lumber companies provide machinery, transportation, and money; affluent people everywhere create the demand; and homesteaders in the forests are all too willing to sell their timber rights and work as lumberjacks. In time they would raze the forests and jungles anyway for agriculture. Those who believe that consumption of resources is the big threat should count all affluent people. All of these judgment calls are too subjective: let us count everybody.

If you disagree, do not be too concerned. Whom to count is less important than it might seem; to a first approximation it does not matter. If we suppose that the "bad guys" who put us at risk are a fraction B of the population, B_p in the past and B_f in the future, then this would alter Equation 10 to read

$$G_m = \cfrac{1}{1 + \cfrac{B_f \times Z_f}{B_p \times Z_p}}$$

If we have no compelling reason to think $B_p < B_f$ or $B_p > B_f$, then the statistically indifferent assumption is $B_p = B_f$. Then the B factors in the above equation cancel and restore Equation 10. The harm that bad guys did in the past has strengthened us proportionately against the harm their descendents will do in the future. Recall the discussion of Jays and Kays in Section 3.1.

Conventional wisdom says that we in the developed world are to blame for almost all the stress on our life-support environment since we consume far more goods per capita than the world's poor. It is fashionable to resent lavish consumers and sympathize with the hordes of poor in the underdeveloped world. However, a fellow who drives a Hummer stresses our environment no more than a typical subsistence farmer in Amazonia. Whereas the driver's extravagance is immediate, the farm family's damage is delayed a couple of generations. They have a high fertility rate and produce myriad grandchildren who mature, raze more jungle to feed their families, and exterminate species in the process. In the subject at hand, the survivability of our entire species, that delay is relatively brief and matters little. Affluent folk are the looters, and subsistence farmers are the locusts. All do damage; so again, let us count everybody.

Let us return to the task of finding a formula for Z, the cum-risk in Equation 10. We digressed after deciding that the annual increment ΔZ must be proportional to population p. However, p cannot stand alone because that would imply that modern man and ancient man have equal ability to consummate extinction. As a metaphor compare Z to the shot count V at the Four-Day Shooting Gallery discussed in Section 1.6. The human race is a distant target, and every person alive takes one shot at it each year. Trouble is, their guns are growing more accurate, and the rounds contain explosives with proximity fuses. Later shots must carry far more statistical weight in order to retain the balance that we call statistical indifference.

In other words, the threat is proportional to the power and expertise people have to commit extinction either deliberately or by accident. This capability must be some measure U of hazardous industrial/technical/scientific development; call it haz-dev for short. Then the annual increment in cum-risk is the product of population times haz-dev:

$$\Delta Z = p \times U$$

Finding a realistic formula for U is the tricky subject of the following section.

4.2 HAZARDOUS DEVELOPMENT

To obtain a formula for haz-dev U we can modify a formula for the world economy, E, expressed perhaps as the Gross World Product in trillions of dollars per year. Johansen and Sornette [31] provided the formula. Their Equation 16 gives the annual increase ΔE. Modifying their notation $(L \to p, B \to C, A \to E)$ to match ours, we have

$$\Delta E = C \times p \times E$$

where C is a constant of proportionality. World population p is a factor because it is the pool of potential innovators and suppliers of labor and materials. Finally E itself is a factor because old productivity leverages new. The steam engine enabled power tools and machine shops, which then enabled all the things we manufacture including better steam engines and power tools.

Here population has arisen in a context different from the one above, which once again raises the question of whom to count. And once again my choice is to count everybody. The world's poor contribute by providing cheap manual labor for everything from mining raw material to assembling the world's machines and instruments. Without them we could not buy a personal computer with a single week's wages. With fewer computers the world would run at a slower pace and be safer for it. Moreover, Third World nations sell off their natural resources at an unsustainable pace, which keeps the First World running ever faster.

Next let us modify the equation above to obtain a formula for U. Like E, yesterday's technology enables today's. In other words, haz-dev exhibits positive feedback, and so a first attempt would simply substitute U for E:

$$\Delta U \stackrel{?}{=} C \times p \times U$$

However, the feedback for U is usually not as strong as it is for E, and so we shall use a fractional power of U:

$$\Delta U = C \times p \times U^{\mu} \tag{11}$$

Two reasons for this reduction both stem from the fact that most man-made hazards involve cutting-edge technologies like genetic engineering, robotics, or pharmacology. First, an ever decreasing fraction of the population has the time, patience and intellect to acquire these ultimate skills. Second, in narrow hazardous specialties there is less leverage among them. Robotics, pharmacology, and genetic engineering do little for one another. An exception is computer technology, a powerful lever for everything.

Physicist Derek Price [32] was the first scientist to do research on scientific progress in general. This field has grown and now has its own journal, *Scientometrics*. Price found that narrow measures of progress exhibit less positive feedback than broader measures. For example, in his time (1963) the number of "important" discoveries doubled every 20 years, while the total number of engineers in the United States doubled every 10.

Some man-made hazards are not so high-tech, for example greenhouse gases and world travel. These gases threaten our climate, and world travel spreads pathogens.

Travel also homogenizes the world, thus reducing the chance that some remote population will have a quirk that makes it immune to the ultimate hazard. For these low-tech threats the exponent μ should be close to 1.0, which changes the whole character of the solution so that U grows exponentially; see Table K in Appendix K. So perhaps we should disaggregate man-made hazards into two categories, one for high tech and the other for low, just as the impresario did for stage productions as described in Section 3.2. Unfortunately, we would pick up a couple more parameters to be evaluated, and we have not enough reliable data to evaluate them all. Besides, our quantities are too vaguely defined to justify the extra complexity. So let us press on and use a single composite value of μ to be determined. It will produce a single composite cum-risk Z, which then represents all man-made hazards.

Appendix K solves Equation 11 with the result,

$$U = (X - X_0)^{\omega}, \quad \text{where } \omega = \left.\frac{1}{1 - \mu}\right\} \atop \text{providing } \mu < 1.0 \qquad (12)$$

In this equation, X denotes population-time, pop-time for short. This is the total number of people-years ever lived by everybody starting from the dawn of humankind, the extent of humanity. It is defined in the same sense that man-hours or person-days gauge the amount of labor to do a job. Each year every person alive adds one more person-year to the X tally. When somebody dies at age A, his life has added A people-years to X. During 2005 world pop-time increased by $\Delta X = 6.5$ billion because that was the world population in that year. The grand total for all past human pop-time is about $X_p = 1.7$ trillion people-years, estimated by adding the world's population for every year since the dawn of our species using estimates from the U.S. Census Bureau [33]. Our ignorance of ancient and prehistoric populations causes little error in X because the populations were so small then.

In Equations 12, X_0 represents the pop-time at some historical tipping point or paradigm shift when man-made hazards began to surge.

Pop-time X will be the main variable throughout the rest of this chapter. Therefore, we need conversions between X and dates, which appear in Table 6. The dates and populations are those used by the economist J. Bradford DeLong [34] rather than the U.S. Census Bureau. This maintains consistency with more of DeLong's data used later. The differences are minor; for example, we get $X(2005) = 2.0$ trillion people years instead of 1.7. Throughout the second millennium pop-times are numerically close to the corresponding dates, which can lead to confusion. Also note that almost half of all the people-years ever lived occur in the second millennium.

Returning to Equation 12, let us determine X_0 and μ by fitting the equation to existing statistical data that use four different proxies as measures of haz-dev:

- the yearly number of United States patents [35]
- the yearly number of pages in *Nature* magazine [36]
- the yearly number of papers published in natural sciences and engineering (NSE) [37]
- an estimate of gross world product (GWP), the sum of all GDPs worldwide [34].

Table 6. Population p and pop-time X at various dates, X expressed in billions of people-years, BPY.

Population X			Population X		
Date	Millions	BPY	Date	Millions	BPY
−300,000	1	247	1300	360	1,310
−25,000	3.3	750	1340	370	1,324
−10,000	4.0	805	1400	350	1,346
−8000	4.5	813	1500	425	1,384
−5000	5.0	828	1600	545	1,433
−4000	7.0	834	1650	545	1,460
−3000	14	843	1700	610	1,489
−2000	27	863	1750	720	1,522
−1600	36	875	1800	900	1,562
−1000	50	901	1850	1,200	1,614
−800	68	912	1875	1,325	1,646
−500	100	937	1900	1,625	1,682
−400	123	948	1920	1,813	1,717
−200	150	975	1925	1,898	1,726
1	170	1,008	1930	1,987	1,736
14	171	1,010	1940	2,213	1,757
200	190	1,043	1950	2,516	1,780
350	190	1,072	1955	2,760	1,793
400	190	1,081	1960	3,020	1,808
500	195	1,101	1965	3,336	1,824
600	200	1,120	1970	3,698	1,841
700	210	1,141	1975	4,079	1,861
800	220	1,162	1980	4,448	1,882
900	242	1,185	1985	4,851	1,905
1000	265	1,211	1990	5,292	1,930
1100	320	1,240	1995	5,761	1,958
1200	360	1,274	2000	6,272	1,988
1250	360	1,292	2005	6,783	2,020

Figures 19 through 22 below show the four fits to Equation 12.

Table 7 summarizes the estimates of X_0, ω, and μ. The last column in the table "quality" refers to the amount of scatter in the data points.

The last proxy in this table, GWP, has $\mu = 1.0$. This confirms that the full feedback in the equation $\Delta U = C \times p \times U$ (above) does happen sometimes. Equation 12 does not apply to this case. Instead, the solution takes a unique form and grows exponentially as shown in Table K, Appendix K. In this example U doubles every 66 billion people-years, which would be 10 years with the present population, 6.6 billion people. The scales on Figure 22 make the exponential process appear as a straight line.

Recall Price's observation that narrow measures of progress exhibit less positive feedback than broader measures. This holds throughout Table 7. The broadest measure of all is GWP, which has $\mu = 1.0$ as discussed above. The second broadest is patents with $\mu = 0.52$, and third is the tally of papers in natural science and engineering with 0.17. Finally, *Nature* magazine has the toughest standards for acceptance. It is the publication most likely to announce an exciting new scientific breakthrough such as the first laser or the double-helix structure of DNA. Therefore, the scope of its individual papers (but not the magazine as a whole) is the narrowest of all, and indeed $\mu = 0$. (Besides, *Nature*'s management may protect their reputation by progressively tightening acceptance standards and thus limiting growth.)

For our purpose *Nature* is too high-tech, and GWP too low. The two remaining sets of data, patents and papers in NSE, are more what we want. The best estimate for

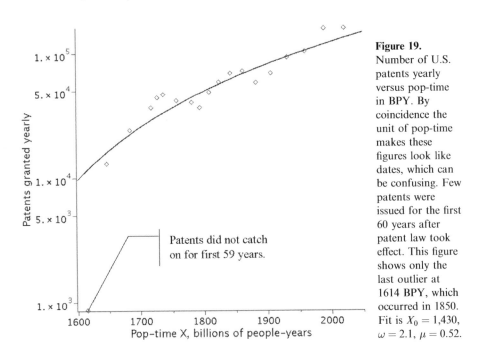

Figure 19. Number of U.S. patents yearly versus pop-time in BPY. By coincidence the unit of pop-time makes these figures look like dates, which can be confusing. Few patents were issued for the first 60 years after patent law took effect. This figure shows only the last outlier at 1614 BPY, which occurred in 1850. Fit is $X_0 = 1,430$, $\omega = 2.1$, $\mu = 0.52$.

Patents did not catch on for first 59 years.

Patents granted yearly

$1. \times 10^5$

$5. \times 10^4$

$1. \times 10^4$

$5. \times 10^3$

$1. \times 10^3$

1600   1700   1800   1900   2000

Pop-time X, billions of people-years

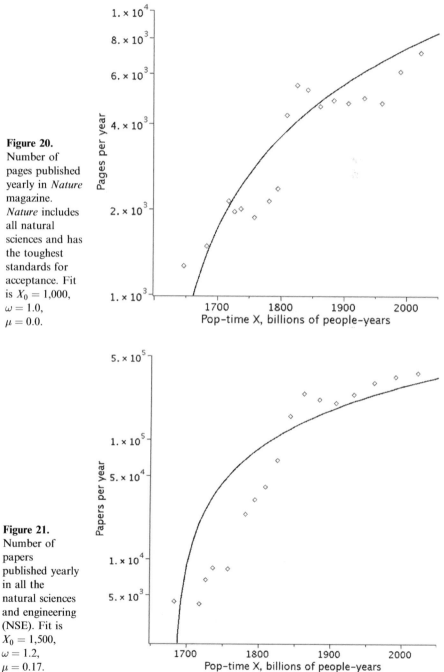

Figure 20. Number of pages published yearly in *Nature* magazine. *Nature* includes all natural sciences and has the toughest standards for acceptance. Fit is $X_0 = 1,000$, $\omega = 1.0$, $\mu = 0.0$.

Figure 21. Number of papers published yearly in all the natural sciences and engineering (NSE). Fit is $X_0 = 1,500$, $\omega = 1.2$, $\mu = 0.17$.

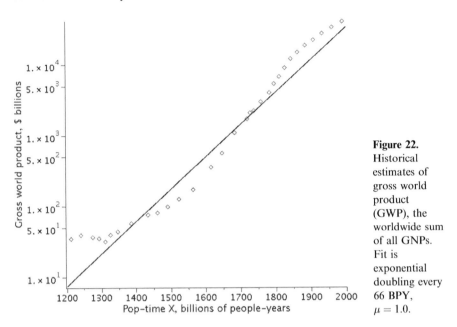

Figure 22. Historical estimates of gross world product (GWP), the worldwide sum of all GNPs. Fit is exponential doubling every 66 BPY, $\mu = 1.0$.

ω will be near the average of the two, about $\omega = 1.6$. The other two proxies serve merely to bracket the chosen pair and to show consistency with Price's observation. We shall use the following numbers, which are close to averages from Table 7, but are adjusted slightly to come closer to X_0 in Figure 20 for *Nature*:

$$\omega = 1.6, \ \mu = 0.38, \text{ and } X_0 = 1,400 \text{ BPY during year } 1530 \text{ AD} \tag{13}$$

This date during the early age of exploration seems appropriate, especially since explorations spread pathogens. Columbus had died in 1506, and a century had lapsed

Table 7. Parameters evaluated using four proxies for haz-dev U. X is expressed in billions of people-years.

Proxy	X_0	ω	μ	Quality
U.S. patents issued	1,430	2.1	0.52	good
Nature magazine	1,000	1.0	0.0	poor
Yearly papers in NSE[a]	1,500	1.2	0.17	tolerable
Gross world product (GWP)[b]	Doubles every 66 BPY		1.0	good

[a] Natural sciences and engineering.
[b] Expressed in billions of 1990 $.

since Admiral Zheng He and his great Chinese fleet explored Southeast Asia and parts of Africa. (Had they explored north and east along the Aleutian islands, I might be writing this book in a Chinese dialect.) Johann Gutenberg's movable type had been in use for almost a century. The first half of the 16th Century was the time of the spinning wheel, Copernicus' discovery that the solar system is heliocentric, the first pocket watch, and the theory of complex numbers. This was the time of the most famous European explorers, mostly Spanish: Ponce de León, Balboa, Magellan, de Soto, de Coronado, Cabeza de Vaca, and de Orellana.

As a reality check let us compare risks in three different years using the tentative parameters in the equations above. Relative risks are easier to comprehend than absolute, so let us compare risks ΔZ during 1900, 1970, and 2005 using the equation $\Delta Z = p \times U$ from Section 4.1 and Equation 12 for U. Let M (for modern) denote pop-time relative to X_0:

$$M = X - X_0$$

Then the ratio for any pair of years is

$$\frac{\Delta Z_1}{\Delta Z_2} = \frac{p_1}{p_2} \times \left(\frac{M_1}{M_2}\right)^\omega$$

Table 8 supplies excerpts from Table 6 to use in this ratio equation for the years 1900, 1970, and 2005.

From Table 8, the extinction risk in 1900 was 6.6% of that in 2005. The possibilities for self-extinction in 1900 are quite far-fetched. Wind-blown agricultural dust has always crossed oceans and could conceivably spread deadly microbes to isolated islanders who would otherwise survive. Or a sea mammal infected by a dirty harpoon could spread a deadly epidemic throughout the world's oceans and infect people wherever sea mammals are hunted. Or both these calamities could happen at the same time and together exterminate both continental people and islanders. There are surely more hazards one could conceive, but regardless it seems like the risk in 1900 ought to be <5%, which would indicate 20 or more hazards in 2005 for every one of comparable severity in 1900. We cannot expect a subjective guesstimate like this to agree better than a factor of two or three with real data from Tables 7 and 8 simply because risks vary chaotically from year to year and decade to decade. Let us accordingly accept <5% as adequate agreement with 6.6% from Table 8.

Table 8. Quantities for comparing yearly risks to 2005. (TPY means tera-people-years, tera = trillion.)

Year	Population p, billions	Pop-time X in TPY	Modern pop-time M, TPY	Risk relative to 2005, %
1900	1.6	1.68	0.28	6.6
1970	3.7	1.84	0.44	39
2005	6.8	2.02	0.62	100

Comparing hazards in 1970 to 2005, Table 8 shows only 39% of the latter. This seems a bit too small, perhaps because it is scary to think that hazards have tripled during my adult life. I had originally selected these two dates as an interval over which the earlier hazard was about half the later. In any case, the comparison 39% to 50% is adequate for checking the reality of something that varies chaotically.

My judgment for comparing 1970 and 2005 was based on historical summaries of the last four decades, but reasonable opinions may differ. You may make your own assessment by studying the hazards listed in Table 9 below and adding your own list to mine.

Table 9. Hazards during 1970 compared to 2005.

Risks during 1970 declining later	• peak increase in world population • threat of nuclear exchange, possible nuclear winter • **Green revolution, consequences unknown**
Risks worsening from 1970 to 2005	• carbon dioxide emission • **worldwide homogenization by commerce and travel** • impossibility of quarantine: air travel, illegal immigration, smuggling, refugees • **huge private fortunes, owners' intentions unknown** • Moore's law, computer capability doubling every 1–3 years • **pharmacology (remember thalidomide)** • high-power microwave emissions that reveal our existence
Risks during 2005 but not during 1970	• genetic engineering • sophisticated robotics • information technology: Internet, Google, and so on

Most of these hazards are familiar, but perhaps not the four in boldface. They illustrate types of surprises that may blindside us:

- Green revolution—Norman Borlaug *et al.* fed more than a billion people who had been chronically hungry by developing and distributing new varieties of grain with higher yield and resistance to disease. Nothing bad has happened, yet, but we didn't know that when the revolution began. Subsistence farmers might have multiplied rapidly, razed the forests, and triggered climate change. Or his tampering with genetics might have caused a disastrous side-effect.
- Private fortunes—The late Sam Walton of Arkansas founded the Wal-Mart chain of stores, which became the world's biggest retailer. What if he (or later his heirs) had "heard" orders from God to end the human race? If they spent $80 billion, the project might have succeeded.

- Homogenization—Worldwide commerce and travel has homogenized cultures, diets, drugs, and so on. This reduces the chance that some isolated population will have some quirk or custom that saves them from the final calamity.
- Pharmacology—Suppose that almost everybody takes a new wonder drug not suspecting its fatal side-effect. This after-effect has a very long latency; nobody suspects anything until it is too late. Then everybody is sterile, or maybe they turn into crazed killers who prowl the land looking for victims.

4.3   PREDICTOR FORMULATION

As shown in Appendix K, Equation K-9, the cum-risk is

$$Z = M^{(\omega+1)} \quad \text{where } M = X - X_0$$

Appendix K also splits Z into past and future, Equation K-10, to evaluate GSP for man-made risks, Equation 10, with the result in Equation K-11:

$$G_m(M_f) = \left(\frac{1}{1 + M_f/M_p}\right)^{(\omega+1)q} \tag{14}$$

where $\omega + 1 = 2.6$ ($\mu = 0.38$) from Equation 13 in Section 4.2;
$M_p(2009) = X(2009) - X_0$ in billions of people-years (BPY);
$X(2009) = 2050$ BPY from DeLong's data, Table 6;
$X_0 = X(1530) = 1{,}400$ from Equation 13, date from Table 6;
hence, $M_p(2009) = 650$ BPY.

These are all the quantities required to evaluate Equation 14 except for q. We must deal with q separately for the two cases: survival of civilization in Section 4.4, and survival of the human race in Section 4.5.

This completes the hard part of the calculation. Just a few simple steps remain before we can obtain our formula for human survivability. First, let us restore the factor for natural hazards found in Section 4.1, namely $G_n = 1/(1 + T_f/T_p)$, to display the complete predictor, namely $G(f\,|\,p) = G_n^{1-q} \times G_m^q$ also from Section 4.1, for the case in which the exponent q is known:

$$G(f\,|\,p,q) = G_n^{1-q} \times G_m^q = \left(\frac{1}{1 + T_f/T_p}\right)^{1-q} \times \left(\frac{1}{1 + M_f/M_p}\right)^{q(\omega+1)}$$

As before, the notation $(f\,|\,p,q)$ denotes the probability of the future f (namely T_f and M_f) given the past p (T_p and M_p) and also given a value for q, the relative severity of man-made hazards. Recall that time is the appropriate cum-risk for natural hazards (because such hazards are as likely to strike one year as another). The second factor in this equation represents man-made hazards, which was dormant for most of human history but is now accelerating rapidly.

#   #   #

The equation above suggests a paradox, and Planet Zanj gives us the perfect paradigm. It is home to a humanoid species with accelerating technology, exactly like ours. However, that species is fortunate enough to have no natural hazards—neither ice ages, bolides, planet-quakes, volcanoes, nor killer typhoons. For them $q = 1.0$ in the equation. This leaves only the second factor, the hazard of self-extinction, and that hazard is maximum with $q = 1.0$, several times greater than ours. Comparing our situation to the Zanjians, it appears that the mere existence of natural hazards on Earth confers major protection from artificial hazards. In our case T_p is a very long time, and so we can ignore the natural hazards on a time-scale of centuries, as discussed in the next two sections. Despite this, the natural hazards seem to protect us somehow! How can this possibly be?

To resolve this paradox, realize that the Zanjians cannot justify omission of natural hazards merely because they have never experienced any. In fact, their inexperience may leave them all the more vulnerable to their first exporsure. Or maybe Zanj is the proverbial powder keg. Perhaps its crust is under extreme stress from shrinkage underneath. The first bolide ever, a modest hundred meters in diameter, may shatter the whole crust and release poison gases.

Also recall that probabilties depend on our knowledge of the process in question. If you are betting on a horse race, your assessment of the odds may change suddenly and radically if you are lucky enough to overhear what the trainers are saying. But nothing sudden happened to the actual physical conditions of the horses or their jockeys. Only your knowledge of them has changed suddenly.

Likewise, the inexperienced Zanjians may make a poor assessment of the statistical weights for q in the equation above. By contrast, after enduring ice ages, watching volcanoes, and generally experiencing nature's fury, we can assess risks better than they can. However, this assessment is a minor adjustment cmpared to complete omission of natural hazards.

#   #   #

The equation above is not quite our final result. The exponent q expresses the relative potency of the natural and man-made hazards. The former gets the greater exponent because nature has many times demonstrated her ability to extinguish species on a massive scale. In other words, q is small. Recall the opening example in Section 3.2 where our impresario evaluated the exponents in his predictor. Using his insider's knowledge, he made a tally of expired shows and disaggregated them into two groups, one for expired popularity and the other for all other causes. Using these two "body counts" and the theory in Appendix G, he had sufficient information to calculate the exponents. To follow his example in the case of human survivability, we would need historical data for expired humanoid species throughout our galaxy.

Lacking these, we could look for existing statistics that would serve as proxies for q. A history of near misses is a possibility: asteroids or bolides that came close to Earth; negligent government officials or military officers during the Cold War who almost triggered a nuclear holocaust; a devastating simian virus that might have jumped to humans but did not. These topics are certainly worth considering for

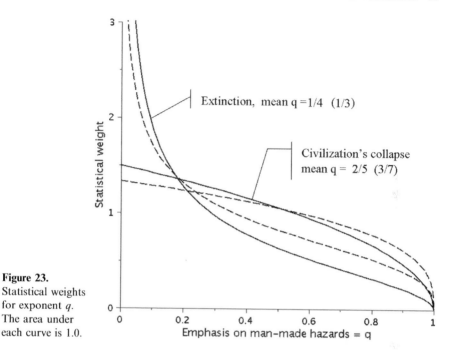

Figure 23.
Statistical weights
for exponent q.
The area under
each curve is 1.0.

future research; they are, however, too much for this treatise. Perhaps you can find a good set.

Therefore our best fallback estimate of GSP, as shown at the end of Section 1.2, is its average over possible values of q. This time we must use an arithmetic average because it represents a logical OR: Hazard 1 may be twice as severe as Hazard 2, OR it may be four times as severe, OR ... The previous average in Equation 8 is geometric because it represents logical AND: The entity must survive both Hazard 1 AND Hazard 2. In what follows, ⟨angle brackets⟩ indicate an OR average.

This change in Equation 14 yields our goal, the formula for human survivability:

$$G(f\,|\,p) = \left\langle \left(\frac{1}{1 + T_f/T_p}\right)^{1-q} \times \left(\frac{1}{1 + M_f/M_p}\right)^{q(\omega+1)} \right\rangle_{0<q<1} \tag{15}$$

The lower limit of the average, $q = 0$, represents the absence of any man-made hazard, while $q = 1$ means the absence of any natural hazard. We must not give these limits equal weight in the average: $q = 1$ is not allowed because the natural hazards will always be with us. Paleontology tells us that nature has the power to extinguish species and has exercised that option frequently. Hence, the statistical weight for $q = 1$ must be zero.

Humans also extinguish species, but only those caught in vulnerable niches like the dodo and the black rhinoceros. We have miserably failed to extinguish many adaptable pests such as rats, mosquitoes, cockroaches, and the destructive invasive species in Australia (rabbits, foxes and cane toads). Humankind belongs in this pesky adaptable category. Despite all man's capacity to wreak destruction and misery, we may lack either the brute force or the worldwide coverage to consummate self-extinction. Moreover, nature is indifferent to who or what survives her violence, while humans shy away from apocalypse as in the nuclear standoff during the Cold War. Hence, $q = 0$ is allowed and its statistical weight is maximum there.

To numerically evaluate Equation 15, we must deduce a formula for statistical weights without the benefit of galactic data or microcosms. In other words we need a formula for prior probability on the interval $0 < q < 1$ in a case where little is known. Jeffreys [21] has addressed the case in which nothing is known. The ideas discussed above can be used to modify his formulas; Appendix L gives the details, and Figure 23 shows the resulting plots. The solid curves are the main ones with maximum statistical indifference. The dashed curves use a slightly contrived variant intended merely to test the sensitivity of the final results.

For extinction the peak weight at $q = 0$ is very sharp because there may be no artificial hazard powerful enough to consummate complete extinction. However, to merely destroy civilization, a disaster need not reach the remotest places. A man-made or man-aggravated hazard (epidemic, bio-warfare) may suffice; accordingly, the maximum statistical weight for civilization is quite mild. For reasons that will be apparent in Section 4.5, we examine civilization's survival first and then species' survival.

4.4 SURVIVABILITY OF CIVILIZATION

Let us estimate the risk of a cataclysm that falls short of extinction. In order to use our formulation, we must pose the question as the survival of some entity that has a rather definite beginning and end. Survival of peacetime does not work because there are always conflicts of various sizes. Survival of a particular nation will not work either: its end may not be a calamity, but rather a merger, a split, or a whole new concept in governance. So let us estimate the survival of human civilization, which may be the most unequivocal example of a lesser calamity.

To qualify as the end of civilization, an event might kill everybody except those in the remotest locations. Humans might survive in the Falkland Islands or Mauritius or perhaps the high Himalayas. Or if islands drown, continental people may survive but with their government, commerce, and urban services in ruins; currency becomes worthless, and survivors abandon cities in their quest for food, water, and sanitation.

To use Equation 15 we need an estimate of T_p, for civilization's past. More than one civilization have appeared independently. For example Maya civilization began in Mesoamerica around AD 200. But that one collapsed and lost its influence during the Spanish conquest. The civilization whose influence spread worldwide and remains today is the one that began about 11,000 years ago, 9000 BC, with agricultural settlements in Mesopotamia. So let us put $T_p = 110$ centuries. Table 6 shows that only 40% of human life (people-years, not the number ever born) occurred before that time.

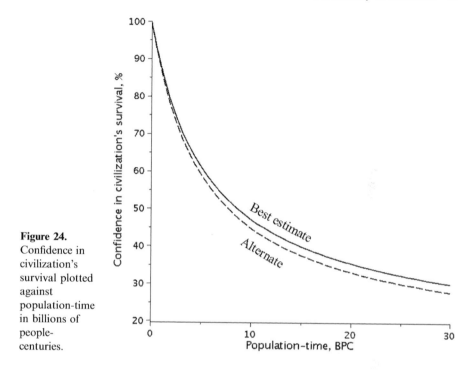

Figure 24. Confidence in civilization's survival plotted against population-time in billions of people-centuries.

There is no reason to calculate survivability beyond a few centuries because so much can happen to alter the mathematical model. This future is short compared with civilization's age. The ratio T_f/T_p in the equation for natural hazards, $G_n = 1/(1 + T_f/T_p)$ from Section 4.1, is on the order of 3%, which is surely smaller than uncertainties in the formulation. Therefore, we can ignore the ratio and put $G_n = 1.0$.

It turns out that this same approximation holds in Section 4.4 for species survival and throughout this treatise, so let us anticipate this result and simplify Equation 15 accordingly:

$$G(M_f \mid M_p) = \left\langle \left(\frac{1}{1 + M_f/M_p} \right)^{q(\omega+1)} \right\rangle_{0<q<1} \tag{16}$$

This is our ultimate predictor, which is very convenient because G now depends only on one independent variable, M_f, which is pop-time after 1530 AD. Thus we can present results as a two-dimensional plot of G against M_f.

Figure 24 shows the final results for civilization's survival. (The dashed curve shows the result using a secondary formula for statistical weights, Figure 23, which serves merely to indicate sensitivity to the uncertainty in those weights.)

A summary result from this figure is that civilization's half-life is only 8.6 billion people-centuries. Let us examine what populations and durations this pop-time might represent:

- If world population stabilizes with a crowd of 12 billion, then the half-life of civilization is about 72 years—almost 3 generations.
- If population drops to a comfortable 4 billion, as in 1974, then the half-life is 215 years—about 8 generations.
- Thus overpopulation may prune as many as 5 generations off the family trees of the many who perish in the cataclysm.

4.5 SURVIVABILITY OF THE HUMAN RACE

Again, there is no reason to calculate survivability beyond a few centuries because so much can happen to alter the model. Contrast this maximum, T_f, with the age of our species, about $T_p = 2,000$ centuries. The ratio T_f/T_p in Equation 15 is on the order of $1/1,000$, which we can certainly ignore. However, one might argue that the effective age of our species is much younger. Evidence suggests that early humanity was severely stressed by unknown forces that reduced our ancestors to a small band of

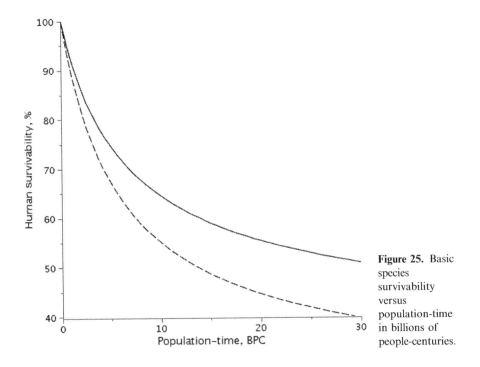

Figure 25. Basic species survivability versus population-time in billions of people-centuries.

survivors, about 700 to 2,700 individuals [38], perhaps 700 centuries ago. Our gene pool has much less variety than those of other primates, which suggests that most human genes were lost in a stressful time leaving only the genes of the few survivors. Such an event is known as a genetic *bottleneck*.

That first group had specialized skills that helped them survive their crisis. But those who died had other skills that the band would need later. This is the same dicey situation that a new business or stage production encounters. The founders have a few important skills and some star performers, but people with other skills are lacking. This parallel makes one wonder whether we should use that time of crisis as the age of modern humans instead of 2,000 centuries. However, the question is moot in our context because the adjusted ratio T_f/T_p is on the order of $1/200$, again negligible, and so the simplified Equation 16 applies again. Since the different ages have no effect, the only significant change from civilization to extinction is the statistical weights used in the average, Figure 23.

Figure 25 shows the results. The half-life for survival occurs at about 30 billion people-centuries (BPC). However, this is much more sensitive to the assumed statistical weight than the case of civilization.

Suppose we had the *Grand Galactic Book of Knowledge* with its database of expired humanoid species. We could pretend to observe them at certain ages and then plot future survivability using *historical* values of pop-time taken from the book. For those sufficiently earthlike, the fit to Figure 25 would be crude but probably satisfactory since we adhered to principles that worked well for microcosms here on Earth.

However, a problem arises when we try to apply this curve to our real future here on Earth because the pop-times are not historical fact but rather an unknown future. Suppose we assume that world population levels off at 10 billion, then each interval of 10 BPC on Figure 25 would represent one century. This would say that the half-life of *Homo sapiens* is only about three more centuries. But this is wrong because the assumed population is not likely to hold that long. According to Figure 24 the chance is 70% that the population will crash in a cataclysm that will kill billions and shut down the hazardous technology. Hence, the real half-life of our species is much longer than three centuries, but only because a lesser apocalyptic event comes to its rescue! The pop-time will not recover and reach the half-life, 30 BPC, for some number of millennia depending on circumstances during the aftermath.

This aftermath will be a time of safety for our species. The cataclysm will relieve human pressure on the environment and let the atmospheric carbon dioxide concentration return to normal. Remote villages will be isolated enough to function as survival colonies should another calamity strike. Lessons from the disaster will not soon be forgotten nor ignored. The event will be a major topic in the history curriculum of every schoolchild for centuries. In effect, the event will immunize humanity against extinction for a very long time, long enough for humanity (or its artificial descendents) to escape Earth and fulfill some greater destiny in the solar system and possibly the galaxy.

Unlike the humanoid data from the book, we do not have actual values of future pop-time to use in the formula. What we have instead are *projections* of future

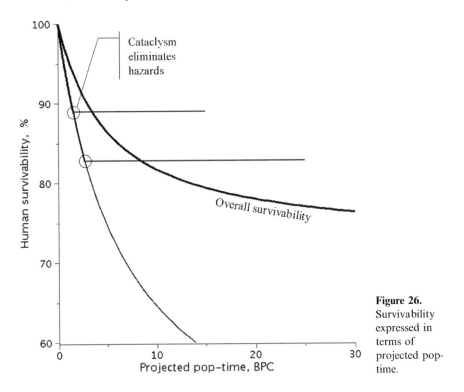

Figure 26. Survivability expressed in terms of projected pop-time.

population and technology made by economists, scientometricists, and related statisticians. The trouble with their projections is that they assume things will continue to run somewhat smoothly just as they have been. They make no allowance for near-extinction events because they do not know how. But we do! We can make a correction that takes cataclysms into account and lets us plot a curve of survival versus projected pop-time, which we can estimate, as opposed to actual future pop-time, which we cannot predict.

Figure 26 shows the general idea. Think of an ordinate (a point on the vertical axis of the graph) as the number of survivors in a big statistical ensemble of humanoid species. The steep curve on the left is a portion of the curve in Figure 25. It shows entities dying off in the absence of any cataclysm. Follow the steep slope downward to a point where a group of species eventually does suffer cataclysms. At that point their risk vanishes, and all survive for the foreseeable future. Meanwhile, their projected pop-time (oblivious to the event) continues like a clock, generating a horizontal line. Further down the steep slope another group suffers cataclysms and generates another horizontal line. These two discrete branches are only illustrations; the real probabilities form a continuum of trajectories that blur into shades of gray. Each trajectory represents a probability, which we know how to combine into a single overall survival probability by using the sum and product rules (logical OR, AND) discussed in Section 1.2. Appendix M gives the details. The bold curve in Figure 26 shows the final overall probability of species survival expressed in terms of projected pop-time.

4.6   SUMMARY AND CURRENT HAZARD RATES

Figure 27 summarizes the two most important results. To the extent that our many assumptions are valid, our species' long-term survivability is 70%, as shown in Appendix M. Recall, however, that "long-term" refers to the aftermath of a collapse during which the population and economy have not fully recovered, and people presumably remember lessons learned from the collapse. Let someone else predict the hazards after that.

Curiously, the number 70% depends neither on ω or M_p, which means that the hazard rate has no effect on the final surviving fraction. If danger to humankind is high, then so is danger to civilization, which rescues our species if it collapses first. If danger to humankind is low, then so is danger to civilization. Therefore, while ω and M_p influence the speed with which things happen, they have no effect on the final

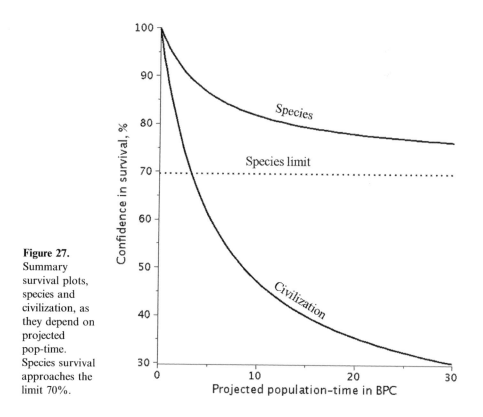

Figure 27. Summary survival plots, species and civilization, as they depend on projected pop-time. Species survival approaches the limit 70%.

outcome. What does affect the outcome is the statistical weights of the exponents for species and civilization, Figure 23.

☺ In the very long term—many centuries—the odds are about 7 to 3 that our species will survive.

☹ It will not be a smooth ride. A near-extinction event may be doomsday for billions of people. However, that cataclysm will probably rescue our species from extinction.

Recall that all results in Figures 24 through 27 stem from Equation 16, where the future depends only on pop-time M_f, not on time or population separately. For example, we might be interested in both survivabilities at 12 billion people-centuries. This pop-time can be lived over a period of 3 centuries in relative comfort at a world population of 4 billion (as in 1974). Lots of nature and biodiversity can be salvaged to keep the biosphere quite pleasant. Or the same pop-time with the same survival prospects can be lived in crowded squalor that stresses the biosphere and its inhabitants while they quarrel over limited resources. Perhaps 12 billion people will endure for 1 century. An extrapolation of census projections peaks at about 10 billion in 2080. From there it could go either way. The fertility decline may continue back to 4 billion or stall and let the population creep up to 12 billion.

$$\# \quad \# \quad \#$$

The current hazard rates are important summary results. They should remain valid for 20 years, more or less. For a big statistical ensemble the hazard rate is simply the percentage of entities that expire per unit time. In our case this means the percentage of Earth-like humanoid species or civilizations that expire per decade.

Appendix N derives the formula for the initial risk, which means the risk at the time of my writing in 2009, $T_f = M_f = 0$. The formula is

$$\Lambda = \frac{\langle q \rangle \times (\omega + 1) \times p}{M_p}$$

Here $\langle q \rangle$ denotes the mean value of the exponent q for the various statistical weights, which are listed in Table 10. In all cases let us use $p = 6.8$ billion people for 2009;

Table 10. Initial hazard rates for making predictions in 2009.

Case	$\langle q \rangle$	Hazard rate, %/decade
Civilization's survival	2/5 = 0.40	9.6
Species' survival, projected pop-time	1/7 = 0.14	3.4
Species' survival, actual pop-time	1/4 = 0.25	6.0

$\omega + 1 = 2.6$ as before, and $M_p = 730$ BPY for 2009. The result is

$$\Lambda_0 = \langle q \rangle \times 24\% \text{ per decade}$$

The first two cases are the important ones that apply to ordinary predictions. The third applies if a cataclysm stops the clock. Note that these hazard rates exceed the rates of some ordinary perils that insurance companies underwrite!

Which is more likely: that your house burns down, or you perish in a global cataclysm? If you live in an ordinary urban house with a fire station at a normal distance, and if you have no implacable enemy, then death in a global disaster is more likely.

Is your fire insurance in effect and up to date?

4.7   BIASES

Let us review a pair of biases that I do not know how to quantify. Each of them may cause our estimates for survival to be too optimistic.

In Figure 22 and Table 7 above we found $\mu = 1.0$ for gross world product, which leads to exponential increase. The growth of hazardous technology U is normally slower than that, according to Price's rule. But nothing prevents an exception. There may be some hazardous activity that is small now, but not for long, because its $\mu > 1$. Some activities that are truly explosive, such as robotics or genetic engineering, may conceivably exhibit $\mu = 1.1$ or 1.2. As shown in Appendix K, the character of the solution changes again in this case, with the result:

$$U = \frac{(-\omega - 1)}{(L - X)^{-\omega}}, \quad \text{where } \omega = -10 \text{ or } -5$$

Note that both $(-\omega)$ and $(-\omega - 1)$ are positive and that $U \to \infty$ as $X \to L$. In other words, parameter L is a drop-dead limit beyond which the entity cannot survive! Maybe this is von Neumann's singularity. Of course U never becomes truly infinite; it only indicates that our mathematical model fails due to dangerous unprecedented events. However, the risk of a near-extinction event is likely to have an earlier L, which would bring on a collapse that rescues the human race as discussed above.

In principle, hazards with $\mu > 1$ should be disaggregated from the composite μ in the manner of Equation 8. In practice I have no idea how to quantify the parameters L and exponent $-\omega$, and so the drop-dead case does not appear anywhere in our results. Yet its possible existence suggests that our results may err on the side of optimism.

$$\#\qquad\#\qquad\#$$

The second optimistic bias pertains to obsolescence. Our mathematical model based on the principle of indifference is not supposed to apply to entities that die of old age. However, no entity is completely ageless; there is always a bit of residual obsolescence. As we saw in Chapter 2, in the case of long-running stage productions in London the last 15% of survivors die off faster than our formula's estimate. Likewise we saw a smoother, lesser decline for long-running business firms. Our predictor makes no allowance for humanity's declining survival fitness, but in fact our species fitness is declining. Modern medicine and social services are keeping people alive and reproducing, many of whom either could not have survived in a prior century, or at least they would have been unable to find a spouse and support a family.

5

Apocalypse how?

The probability of humankind's long-term survival is encouragingly high, roughly 70%. This implies survival long enough to colonize the solar system and perhaps the galaxy either with *Homo sapiens* or some sort of conscious artificial creatures that we regard as our intellectual descendants (cyborgs, androids, whatever).

However, the high probability of survival does not imply smooth sailing. The chances are about 50–50 that some apocalyptic event will decimate mankind within the lifetime of today's newborns. That will stop the frantic pace of technology and development and make the aftermath a very safe time in which the biosphere recovers from human devastation, although perhaps with a changed climate. Meanwhile, the present is a very dangerous time with almost 4% probability of extinction each decade. The probability of a lesser apocalypse is about 10% per decade.

The ultimate calamity, extinction, must reach the mountains of Tibet, the jungles of Amazonia, underground malls and subways of great cities, the Falkland Islands and the remotest islets in the South Pacific. It must be pervasive in the extreme, leaving so few survivors that they cannot find each other and assemble a tribe that has all the essentials for reproduction: a mate, perhaps a midwife, and enough people to comprise a viable breeding stock, perhaps eighty [39].

Nothing can save us from nature's most potent disasters, the classic example being collision with a big asteroid. However, for other hazards, survival colonies reduce the probability of extinction. Remote outposts built for other purposes also serve as unintentional survival colonies.

Teams living in Antarctica comprise a survival colony. At McMurdo Station the few who stay through winter number about two hundred. They have contingency

McMurdo Station, Antarctica. Discovery Hut, Scott's starting point on his first attempt to reach the South Pole in 1902, is in the foreground on the left, Observation Hill is on the right, and McMurdo is between the two. There are about 40 large buildings, many large fuel storage tanks, and countless rows of piled up cargo in McMurdo.

plans and supplies for long isolation. Smaller permanent bases bring the total to about a thousand. Some bases possess the means to escape unassisted before their supplies run out. However, they are not completely immune to global catastrophe. Travel during the summer might bring in an epidemic. An event that cools Earth (such as an asteroid strike or extreme volcanism) during the austral winter would likely kill them all [40].

Men outnumber women about two to one in these teams. If Antarctic teams are the sole survivors of some apocalyptic event, their numbers provide more than adequate genetic diversity for long-term survival of our species. In the first generation, polyandry would improve their gene pool and the men's behavior.

The fact that remote outposts exist now is no guarantee they will be there when survival depends on them. Sooner or later these outposts will be abandoned, most likely after losing their funds to competing demands. This leaves humankind vulnerable to any hazard that would spare only the location of the abandoned outpost. Let us ask Gott's predictor how long the Antarctic research stations will survive. Serious Antarctic studies began at a rather definite time, the International Geophysical Year of 1957, 52 years ago (from 2009), so Equations 3 in Section 1.4 tell us,

$$\text{Duration of Antarctic crews} > 6 \text{ years with } 90\% \text{ confidence}$$

$$> 52 \text{ years with } 50\%.$$

5.1 SCENARIOS FOR EXTINCTION

Population-dependent threats are all too obvious and worsening. Since 1950 world population has surged from 2.5 billion people (BP) to 6.3 BP, a factor of 2.5. New epidemics appear so often that we give them initials instead of names, such as AIDS

(Acquired Immune Deficiency Syndrome), SARS (Severe Acute Respiratory Syndrome), and BSE (Bovine Spongiform Encephalopathy). Man and nature may cooperate in our final demise. Nature creates the deadly disease, and then human activity spreads it. It is well known that dust carries microbes across oceans reaching the remotest islands and the highest peaks. Nature has always produced dust, but humans make more of it and add different sorts of contaminants.

Our mobility may spread disease before quarantine can be enforced. In the 14th century the horrific Black Death never threatened extinction, because it was confined to Eurasia. People were more mobile in 1918 when Spanish influenza hit. That disease spread worldwide and killed 20 to 40 million, about 2% of the world's population. A comparable epidemic today would spread much faster and spare fewer areas if any. It is impossible to quarantine a continent in view of all the varied modes of transportation and the large number of travelers. Besides people taking legitimate business and personal trips, there are covert smugglers, illegal immigrants, and boat people.

Carbon dioxide may be another hazard. In about 2050 its atmospheric concentration will approach twice its preindustrial value. This concentration has been very high in the distant past, but humankind and agriculture were not yet present to interact and cause hazards that are not yet anticipated.

Safeguards against well-known hazards are already in place. We carefully monitor the atmospheric concentration of greenhouse gases, we continually update numerical models of climate change, and the Spaceguard program tracks those asteroids in orbits that may possibly collide with Earth. However, we should not forget that safeguards fail quite regularly. Since serious threats occur rarely, our guardians have plenty of time to grow lazy and complacent. The Federal Emergency Management Agency (FEMA), for example, was totally inept when Hurricane Katrina struck New Orleans. An airport security system was in operation on September 11, 2001 when terrorists hijacked four big passenger aircraft and used them for kamikaze attacks. The Securities and Exchange Commission (SEC) is supposed to protect investors from fraud. Yet Bernard Madoff, a respected financier on Wall Street, was able to run a Ponzi scheme that cost investors $50 billion despite financier Harry Markopolos claiming that he tried for nine years to alert the SEC. And so disasters continue.

The doomsday attack may well be deliberate. The culprit may be the proverbial mad scientist or a demented trillionaire who spares no expense. We should not forget that offense has a huge systemic advantage over defense. The attacker can spend a long time to find one optimum opportunity. By contrast, the defense must watch all possible vulnerable points all the time.

Before long, new hazards will emerge from the realms of science fiction and take their place on the list of real concerns. Robotics is a big concern, genetic engineering another. A more far-fetched hazard is the increasing number of high-power beams such as radar and lidar that are probing the sky and announcing our location to whatever may be out there.

#   #   #

In his book *Our Final Century* [2] Sir Martin Rees discusses the rapid arrival of new hazards in recent decades without focusing exclusively on complete extinction of our species. John Leslie [12] provides an all-inclusive list of extinction hazards in his introduction to *The End of the World*. In the popular literature Corey S. Powell [41] provides a summary of specific threats. The following is a list of the extinction hazards borrowed from his section headings:

Natural disasters
1. Asteroid impact
2. Gamma-ray burst
3. Collapse of the vacuum
4. Rogue black holes
5. Giant solar flares
6. Reversal of Earth's magnetic field
7. Flood–basalt volcanism
8. Global epidemic

Human-triggered disasters
 9. Global warming
10. Ecosystem collapse
11. Biotech disaster [genetic engineering]
12. Particle accelerator mishap
13. Nanotechnology disaster [self-replicating microbes]
14. Environmental toxins

Willful self-destruction
15. Global war
16. Robots take over
17. Mass insanity

A greater force is directed against us
18. Alien invasion
19. Divine intervention
20. Someone wakes up and realizes it was all a dream

In number 17, mass insanity, Powell refers to worsening statistics for mental health. However, this heading could also refer to a mental state induced by a new wonder drug that almost everybody takes before its unexpected deadly side-effect appears. Perhaps 70% of humankind lose their minds, go on a murderous rampage, and kill the other 30%.

The results of Chapter 4 allow a major revision to this list. The disasters in which humans play no role, numbers one through seven, are negligible, simply because our species has been exposed to them for a very long time and has thus acquired a very long track record for survival. The same reasoning applies to 18, 19 and 20. Number 12 seems too far-fetched.

War, number 15, is not significant demographically unless it is part of some complex scenario involving other hazards. If nuclear war occurs, the blasts will mostly be confined to the northern continents. From that origin nuclear winter would be bounded by the latitudinal bands that dominate air circulation. Even if the tropical Hadley cells merge, the temperate Ferrel cell from 30°S to 60°S would stay mostly intact. Inhabitants of Falkland Islands, Tasmania, New Zealand's South Island, and Tierra del Fuego would survive. Philosopher Quentin Smith has independently made the same observation [42].

In no special order, the serious hazards that remain are the following:

- Global epidemic
- Nanotechnology disaster
- Global warming
- Environmental toxins
- Ecosystem collapse
- Robots take over
- Biotech disaster
- Mass insanity

These hazards are all so well known and widely publicized that there is no need for yet another review. Besides, the hazard that eventually kills us will probably not be any of those listed above, nor will it be monitored or well publicized. Rather it will be something bizarre that blindsides us because nobody has thought of it, or if they did, they never took it seriously. To stress this point, the following eight scenarios emphasize the offbeat and unexpected. Fictional details in some of them may help you visualize a sequence of events and decide whether it is truly plausible. In no special order the scenarios are the following:

1. Mutant phytoplankton
2. Coincidence and press/pulse
3. Latent killer
4. Pharmacology
5. Runaway greenhouse effect
6. Instability
7. Self-sufficient, self-replicating robotic species
8. Conspiracy

This set serves as a foil to the balanced discussions by Rees, Leslie, Powell, and others. The choice of eight examples is purely arbitrary. Their purpose is not orderly coverage but merely examples that indicate a range of possibilities. The actual number of such complex unorthodox scenarios is virtually infinite, hence the high risk.

1. *Mutant phytoplankton* Trace gases in the atmosphere can have a big effect as demonstrated by chlorofluorocarbons decomposing the ozone layer. The source of that gas was artificial, and so governments were able, with difficulty, to control the emission. But what if the source were mutant plant life? Earth's plants have

already changed our atmosphere in a big way: they gave us oxygen. The next change may be toxic to humans. If poison gas comes from a land species, we will identify the culprit and exterminate it at all cost before it spreads too far. Since the target is stationary, we would win that battle even against the most aggressive species. But what if the peril is phytoplankton or perhaps kelp or some other seaweed? We cannot treat the world's oceans with herbicide, so that would be our doom.

Nature has warned us with the so-called *red tides* (back cover). They are not really tides but rather algal blooms in shallow seawater. The water turns reddish-brown with plankton, typically *dinoflagellates* (which means "terrifying flagellates"). The plankton poison shellfish, which in turn poison people who eat them. Occasionally victims die.

A related dinoflagellate is more vicious. *Pfiesteria piscicida* attacks with deadly nerve toxin. It was discovered by JoAnn Burkholder, a biologist at North Carolina State University [43]. In 1991 it killed as many as a billion fish in warm shallow estuaries nearby. After Burkholder sampled a toxic tank using gloves and the usual precautions, she suffered nausea, burning eyes, cramps and loss of memory. Five of her colleagues were seriously affected and suffered severe short-term memory loss. In one case, the victim had not been near the tanks. He was merely downstream in the ventilation system and had probably inhaled toxic aerosols.

Marine aerosols would probably not be a threat to people in the highest inland mountains such as the Himalayas, but toxic gas emission is a possibility. One estimate [44] states that living plants emit 400,000 tons of organic volatiles into the air, even some with metals in their molecular structure. Other estimates say millions of tons are emitted [45], the differences perhaps depending on what gases and vapors are included.

It is unlikely that a toxic-gas mutation can happen as a purely natural phenomenon. If that were possible, it most likely would have happened already, probably millions of years ago. Well, maybe it did. In the past 540 million years there have been five major events that killed more than half the animal species. Lesser mass extinctions that killed at least 10% bring the total to 26. Although a bolide strike has been implicated in the most recent event, the one that killed the dinosaurs 65 million years ago, other hazards may account for some of the other mass extinctions. A mutant species of phytoplankton is a remote possibility.

Chapter 4 showed that extinction risk from purely natural events is negligible compared to hazards that involve human activity. Genetic engineers and/or polluters may indeed create the deadly mutant by accident. Radioactive effluents certainly cause mutations, and one of them may be the eventual culprit. Once made, the mutant organism reproduces freely and propagates on its own throughout the oceans.

The only possible defense is risky and unlikely to happen, a survival habitat for the privileged few. It might be an airtight artificial biosphere something like Biosphere 2, the first attempt in Arizona, but bigger this time and with all the bugs worked out. The refugees must escape eventually before crucial supplies are exhausted, or equipment wears out. Perhaps there is a way the survivors can make their descendants immune by vaccination, selective breeding, or whatever, but it all seems too complex to succeed on the first and only attempt.

It seems unlikely that any sophisticated survival habitat will be built and fully maintained prior to its need. Until the need is apparent, the project would suffer from severe lack of urgency and political will. It would be a prime target for budget-cutters who look for projects that produce no conspicuous benefit in the short-term. However, if we wait until the need arises, there will not be much time. Without many years of testing, the habitat will surely fail, as did Biosphere 2.

2. Coincidence and press/pulse In this scenario one mass killer leaves one set of survivors, another leaves a different set. Occurring together, they kill everybody. The dinosaurs' demise may have been just such a double whammy: the bolide strike near Chicxulub, Yucatan, Mexico plus major volcanism at the Deccan Traps in west-central India.

The ultimate calamity will most likely be an improbable coincidence of multiple hazards or failures simply because we develop defenses only against single hazards. Defense against coincidences is impractical because there are too many possible combinations, each of which is very unlikely.

Some studies of past mass extinctions support a so-called *press/pulse* model [46]. *Press* denotes a long-term pressure on the ecosystem, and *pulse* a sudden catastrophe. The model makes good sense because two pulses are unlikely to coincide, and two

Deccan traps in western India The word *traps* derives from the Swedish word *trappa* for stairway. Successive lava flows piled on top of one another in a manner that looks like stair steps.

presses probably leave survivors who manage to adapt. In the dinosaurs' case volcanism might have been the long-term pressure because dust blocks sunlight and volcanic gases pollute the air. The bolide strike at Chicxulub, Yucatan is the prime suspect for the pulse.

#   #   #

As noted above, war by itself is not enough to cause extinction. However, while nuclear winter depopulates the Northern Hemisphere, a coincidental natural event might take out the South. Volcanism is a possibility. It may occur at one or both of the southern tips of the Pacific Ring of Fire in or near Chile and/or New Zealand.

As a third example, suppose climate-monitoring stations in Greenland and the southern Indian Ocean fail simultaneously in a way that masks an Orange Alert. By the time the trouble is fixed, it has become a Red Alert. That's the pulse. Coincidentally there is long-term tension between industrial nations and the underdeveloped world. Leaders in four populous Third World nations assert themselves by refusing to respond to the Red Alert because that would be seen as cooperation with the industrial world. They claim the alert is a political ruse and refuse to shut down their national consumption of fossil fuel. Before the United Nations (U.N.) can enforce the Climate Treaty, it is too late. Earth follows the path of her sister Venus.

3. *Latent killer* This fatal contagion has a very long incubation period, as do BSE and AIDS. By the time its symptoms appear, air travel will already have carried it to the tiniest populated islands and all the remote outposts. Well, maybe not *every* remote settlement, but perhaps a coincidence will strike those that remain. Wherever the epidemic strikes, it will infect everybody during its latency. When symptoms appear, it is already too late for quarantine. This is not too far-fetched: BSE has been known to lie dormant for decades without symptoms. Incidentally, if a microbe kills too quickly, its victims have little time to infect others. Thus, latency serves the microbe's reproductive interest; in other words, natural selection reinforces this trait.

Although diseases kill individuals, they rarely kill their host species. That would be contrary to the microbe's reproductive interest. Therefore, the killer probably will not evolve naturally. But genetic engineers can make it happen either by accident or on purpose. A lone mad scientist is not out of the question. In the words of Richard Posner, "Human extinction is becoming a feasible scientific project" [47].

4. *Pharmacology* This hazard also depends on latency. Imagine that a wildly popular wonder drug has an unknown side-effect: those who take it slowly become sterile, and their offspring are born sterile. In our homogenized world, tourists, anthropologists, missionaries, photographers, and foreign-aid workers carry the drug to the remotest outposts. Nobody has any warning until middle age when grandchildren fail to appear. Nature has warned us with the drug diethylstilbestrol, better known as DES, which is now banned in the United States. It has produced adverse transgenerational effects in the reproductive tract. Thalidomide, which produced defective children, was another warning of this sort of hazard.

On a longer time scale, pharmacology weakens our species by breeding dependency (press). Worse, it tends to breed pathogens that mutate rapidly (as do influenza and HIV [Human Immunodeficiency Virus]). This is the pathogen's survival trick by which it complicates or defeats medical treatment.

Although these hazards could cause the collapse of civilization, they are unlikely to cause extinction because there will always be some group that refuses to take the drug, perhaps a religious sect. To consummate extinction we must postulate something more violent. Maybe people taking the drug will go berserk and roam the land looking for victims to kill. Lurid fiction has zombies turning into murderous cannibals. As fantastic as it seems, we cannot entirely rule out the possibility that a drug might induce a murderous rampage.

5. *Runaway greenhouse effect* Our solar system has already demonstrated this hazard in the case of Venus, Earth's near twin [48]. She has Earth's size and a nearby orbit. Apart from slow rotation and dense atmosphere, Venus may have been much like Earth during infancy before her water evaporated into space. It is remotely possible that volcanism or a bolide strike could flip Earth to this lifeless state, but this hasn't happened in the last billion years. Thus the probability of this natural event is insignificant, as Chapter 4 has shown. However, humankind may find a way to aggravate this natural hazard.

It is not clear how we could set off a cataclysmic event. It might involve methane (natural gas), CH_4, a powerful greenhouse gas. Vast undersea deposits of *methane hydrate* [49] will release methane if warmed and/or if its pressure is relieved. Perhaps our constant quest for hydrocarbon fuel will accidentally release a burp of methane, especially if the hydrate is already unstable as a result of ocean warming or falling sea level. Further warming might release more methane until it becomes a cataclysmic eruption. (Incidentally, methane hydrate is also known as *methane clathrate*, and the eruption is a *clathrate gun*. Never underestimate scientists' ability to create jargon.)

6. *Instability* On the evening of November 9, 1965 an electric power failure blacked out parts of Ontario, Connecticut, Massachusetts, New Hampshire, Rhode Island, Vermont, New York, and New Jersey. In some areas it lasted 12 hours. Twenty-five million people were affected. The cause was eventually traced to human error that happened days before, when a maintenance man incorrectly set a protective relay on a transmission line. In 1881 when Thomas Edison's central power system electrified New York, he could not possibly have imagined such complex unstable connectivity over such a huge area.

Although that power failure was insignificant compared to extinction, it did demonstrate the downside of big complex systems with lots of interconnectivity. They work wonders while they work, but they also tend to have subtle instabilities that wreak havoc when least expected. Seven states and a province are a tiny fraction of the whole world, but if complexity multiplied that much in 84 years, think what it may be like in the next century or two. In particular, the explosive growth of the Internet makes our economy increasingly dependent on its proper functioning.

Cause of electric blackout in Lisbon and half of Portugal, May 9, 2000; well, not this particular stork, but a similar one.

By itself, complexity probably has no means to deliver the *coup de grâce*, but it may be a contributing factor in some complex coincidence.

7. *Self-sufficient, self-replicating robotic species*   Self-replicating robots are not feasible today nor in the near future because they cannot fabricate the integrated-circuit chips they need to make their own brains. Billion-dollar facilities are required for that task. However, someday there may be artificial brains that grow by themselves when bathed in a special nutrient. The process may be something like either crystal growth, or biological growth of brain tissue in an embryo. The brains may be two-dimensional like integrated-circuit chips or three-dimensional like biological brains. A three-dimensional design will have a serious cooling problem, but perhaps that will be solved if coolant channels are an integral part of a crystal-like lattice.

Ultimately the robots will be capable of foraging for their own fuel and supplies and making their own spare parts. The danger is that a self-replicating species will proliferate, overrun the earth, and devour everything. Responsible nations will forbid these species by law. However, in a world with billions of people and nearly 200 nations, what is feasible usually happens somewhere. A group of rogue engineers funded by a rogue trillionaire may develop them in secret. Or, if not in secret, they can just do it quietly in a nation where it is legal. To keep it legal, the trillionaire puts national leaders on his payroll.

These robotic species will come in assorted shapes and sizes, one or more of which may prosper unaided. There will be safeguards against runaway reproduction. For example, a bigger stronger species (not self-replicating) will be programmed to be its "natural enemy". But then hackers may find a way to disable the safeguards. Even without hackers, mutations and mistakes occur, and when the bad one happens, the prolific species overruns Earth.

If the robots are tiny, they will spread worldwide, just as rats have done. They will hitch rides in luggage, boxes of freight and bilges. If the successful robots are too big to spread by accident, then the infestation can be confined to one continent until a human misanthrope smuggles them out. A sub-billionaire can spread the scourge to every habitable continent and island if she can hide her intentions from her hirelings and smugglers.

8. *Conspiracy* Several decades in the future, embryo selection will be available worldwide. Each embryo will be selected for its genetic perfection from a batch of a half dozen or more. Most industrial nations will offer embryo selection free of charge to all disadvantaged prospective parents. Besides the humanitarian aspects, this policy makes financial sense. The cost of selection is more than offset by the reduced need for social services and medical care later in life.

In this scenario a clandestine organization arises, the Secret Eugenics Society (SES). Most of its members began life as designer babies. Many are second-generation selectees. They observe that a majority of citizens worldwide reject genetic services and reproduce the old-fashioned way. SES members view this behavior as child abuse on a grand scale. Why would anybody want their children to have inferior health and intelligence? How can people justify using their mediocre eggs and sperm when selection is readily available? As members interact, they reinforce these views and develop growing contempt for the majority of humankind.

Eventually, the SES regards most of humanity as hopeless. Despite aid, underdeveloped countries remain underdeveloped generation after generation. Overpopulated countries remain overpopulated despite efforts to promote family planning. Subsistence farmers continue to strip the land, kill off species, and exhaust natural resources. "And for what?" the eugenicists ask each other. If impoverished lives had a modicum of quality or any purpose in the big picture, then many SES members would feel a moral compunction to help them, or at least to let them live. But now the Secret Eugenics Society has grown weary of it. They now regard the masses not as real people but rather as an infestation to be exterminated.

By a big majority, the SES decides that the most merciful solution and the only hope for global happiness is to cull humanity. "And why not?" they ask one another. People cull herds of goats, elephants and other animals when their numbers threaten their habitat. Why shouldn't excess humanity fall in the same category? All those people must someday die anyhow, so to SES members it seems reasonable to hasten the inevitable if that is what it takes to make a better world. The SES plans a surprise attack to kill all humanity except for several thousand privileged survivors, which of course include their own membership.

Is this scenario absurdly far-fetched? Probably not. Attitudes and morality change; what was outrageous a few decades ago is routine now. Why should this trend stop? Before 1973 abortion was illegal throughout the United States. During the 1950s college dormitories for women had curfews and were guarded like fortresses against sexually aroused men. Go back a few more decades and alcoholic drinks were forbidden in the United States, a period known as Prohibition, 1920 to 1933. Well into the 19th century slavery was common, and men fought duels. No one can predict the next change in morality.

The end of the world may happen as follows: The Secret Eugenics Society conducts studies, which show that biological weapons are the best choice. The Society can make weapons by genetic engineering, which has become routine and inexpensive. Genetic engineering is on a roll, much like the computer revolution of the 1980s, 1990s, and 2000s. The SES has skilled research biologists and genetic engineers among its members, plus money to hire many more, most of whom are kept unaware of the project's true goal. The SES carefully selects and breeds pathogens to be as lethal and contagious as possible to humans while sparing most plants and livestock. They also develop a vaccine to protect the designated survivors. The pathogen's ability to reproduce is designed to fail after one year, long enough to exterminate humanity plus a few months to spare.

Committees draw up a detailed plan. The big expenses include research and development and a stockpile of weapons. All together they cost hundreds of billions of dollars. But money is not a serious problem; in recent decades private fortunes have been growing at an unprecedented pace. Nearly all SES members are millionaires, a few are billionaires, and one of them is among the first trillionaires.

The SES develops aerosol bombs to dispense the pathogens. The bombs are to be hidden in dark remote corners of public buildings, subways, and the like. Very few people would see them, and the few who do would be deterred by their false labels: "Do not disturb, Property of the Department of Air Quality Control." The bombs are to be planted worldwide with signs in hundreds of languages.

The big problem is testing the plan. The SES sets up an "industrial biotech plant" on an isolated Indonesian island. It has an "accident" and everybody dies except a test group who were vaccinated. But that is only one test; several more are needed. Years and continents must separate the next test from the first so that the world does not suspect a conspiracy. The second test occurs on an island in the Canadian Arctic with a mix of Inuit and Caucasians. The test succeeds, meaning that vaccinated people survive and the others die. But this time a worker escapes and spreads an epidemic among the sparsely populated villages. Information leaks, hired assassins kill those who know too much, bribes are paid, and the SES barely manages to contain the damage.

The membership grows impatient. On a close vote, they decide to forego further testing—big mistake. The next extermination is worldwide. The pathogens kill everybody who was not vaccinated. However, in the Amazon jungle the pathogen does not expire after several months as it is supposed to. It finds a host where it thrives, reproduces, migrates, and mutates (shades of *Jurassic Park*). Soon the vaccine is ineffective, those chosen to survive die like the others, which completes the extinction of humankind.

5.2 WILD CARDS

Wild cards can be either our downfall or salvation.

India and China These nations have two billion people and sizzling economies (prior to the global downturn in 2008). At present they show little interest in big long-term issues like survivability and global warming, but then we cannot expect them to think that way until their living standard improves. This is happening, but so is their consumption and their impact on the biosphere. Will they become world leaders in conservation, survival habitats and related projects? Or will they opt to be uninhibited consumers? Nobody knows. Perhaps our species' survival is at their mercy.

Outrageous private wealth *Forbes* magazine publishes an annual report on the 400 richest Americans. In 2006 for the first time, all 400 were billionaires. In 2004 there were 313 billionaires, compared to 262 in 2003. The five Walton heirs (Wal-Mart stores) have $90 billion among them, and Bill Gates (Microsoft) has $53 billion. At the rate private fortunes are accruing, there will be trillionaires before long. Any one trillionaire would be able to buy weapons of mass destruction. She might possess enough wealth to purchase a nation, rewrite its laws, and take its seat in the United Nations, perhaps in the Security Council. Are all these trillionaires sane? What about their heirs? Does the Federal Bureau of Investigation (FBI) or any intelligence agency keep tabs on them?

Several people have committed horrendous, senseless crimes and claimed that they acted on orders from God. Trillionaires are not immune to such delusions. If and when a trillionaire imagines God's command to end the human race, he may have the will and the power to succeed.

John E. du Pont gave us a small demonstration [50]. He is the great-great-grandson of E. I. du Pont, founder of the chemical company. As one of many heirs, John's fortune totaled a couple of hundred million. He is seriously deranged and sometimes has claimed to be the Dalai Lama (among others). In 1996 he shot and killed wrestler David Schultz for no apparent reason. A jury convicted du Pont of third-degree murder. Okay, so he killed only one, far short of 6.5 billion, but his crime and mental state warn us that the wealthy can have murderous delusions like anybody else. They have the money to play out their delusions, and they are less likely than others to be under supervision.

We think of megafortunes as good capitalism—an incentive to perform, a just reward for a job well done. True, to a point, but gigafortunes far exceed that point, simply because a thrifty shopper can live her whole life in splendid luxury for only a few hundred million. After that, the money is just a thrill that feeds a dangerous craving for power and enriches heirs of unknown sanity. Perhaps governments whose tax laws allow such accumulation should be considered corrupt. In 1965 chief executives of U.S. corporations were paid 24 times the wages of the average worker. By 1989 that ratio had crept up to 71. In 2005 it was 262, an increase exceeding an order of magnitude in only forty years. To quote Francis Bacon: "Money is like muck, not good except it be spread [51]."

Ironically a billionaire or trillionaire may save us when government fails. There is no hope of making human survival a viable political issue since it lacks urgency and emotional appeal. How many politicians are capable of understanding this treatise? What fraction of their constituents? But hope is not lost. The possibility remains that one of those vast fortunes can save humanity. Wealthy landowners in some countries have protected private forests that peasants would raze were the land equitably divided. Billionaire Gordon Moore (of Moore's law) is a major contributor to Conservation International, a group protecting land that would otherwise be plundered. Albert Gore, Nobel Laureate, is a crusader to stop global warming.

Here is a modest proposal to save the world that an ordinary billionaire can sponsor: *Help nature take back the tropics*. From the dawn of our species until the mid-20th century, the tropics were sparsely populated. They were a reliable buffer against the excesses of humankind. The tropics absorbed carbon dioxide, stored carbon, released oxygen and water vapor, retained nutrients, and preserved the genes of millions of unidentified species of unknown (but probably great) future value. Then "progress" came to the tropics with steel tools, insecticides, fungicides, and medicine. The population exploded, and with it the demand for timber and raw materials. Now the great tropical buffer is dwindling.

Perhaps a wealthy preservationist's most effective strategy is to aggravate the tropical conditions that discouraged development in the past. Insects and vermin are high on the list. One possible project would breed and/or import hardier more virulent pests, something aggressive and resistant to insecticides, repellents, and the like. Genetic engineering might play a role. Unlike two-legged forest rangers, the six-legged kind are far more numerous, self-replicating, and respect neither property lines nor national boundaries. Best of all, they work without pay.

Homesteaders would suffer terribly from this preservation project. Depending on the date and the areas affected, hundreds of millions might be displaced, but the billionaire sponsor cares little; she is dedicated to her greater goal. Moreover, she *does* care about the primates that the homesteaders abuse. In the case of Africa, she cares about gorillas and chimpanzees, and in the case of Amazonia, she cares about indigenous people. She views her project as long overdue redress for a half century of abuse and encroachment on their lands. She also realizes that the indigens are humanity's fallback population, which should never be assimilated for reasons discussed in Section 5.7 under *homogenization*.

An alternative project would plant land mines in the main roads leading into the forests and jungles. This might work where the terrain offers plenty of foliage to conceal implanting teams and their equipment. Roads are the bane of preservation. Main arteries branch into modest roads, which in turn branch into dirt roads and driveways until vast areas are open to development, exploitation, and poaching. A billionaire preservationist can probably acquire sophisticated mines and implanting equipment normally reserved for the military. These mines can be activated or deactivated by remote control. If observation posts are feasible, mines can be inactive

Deforestation in Amazonia
as seen from satellite.
Small roads branching
from big ones make this
typical fishbone pattern
(Wikipedia, public
domain).

while sweepers are present, active as soon as they leave. Minesweeping operations are likely to fail for other reasons. The mines are difficult to detect against the natural background of rocks, roots, and irregular terrain. Neither magnetic detection nor acoustic echolocation is likely to work. The detonator's trigger can detect either pressure or magnetic material such as steel. The latter is attractive because it would take out bulldozers and lumber trucks while sparing pedestrians and carts drawn by animals.

Robotics   When a species expires, it usually succumbs to predators or to competition from another species that has moved into its territory and/or ecological niche. Within the biological realm humankind has no such challenge, but in a larger scope that includes robotics and artificial intelligence, we shall soon manufacture our own competition.

Sometime this century we shall see a revolution in robotics similar to the recent one in personal computers. (Perhaps that revolution will produce the first trillionaire.) What guarantees this is a most amazing example of runaway technological growth widely known as Moore's law. In 1965 Gordon Moore noted that the density of transistors on a computer chip had doubled every year. The time slipped to a year and a half, then to two. Sometimes the law takes on a broader interpretation as a doubling of computational power. For rough estimation over a long term, two years is a valid doubling time. That amounts to a thousandfold increase every 20 years, and a millionfold increase during the forty years from 1968 to 2008. Other components have also maintained the pace. Especially noteworthy is the capacity of the hard drive. It went from a few megabytes in 1980 to a terabyte in 2008 by doubling every 1.5 years.

Obviously a physical limit exists, and so the growth must level off at some point like a logistic curve. Moore's law will fail someday when the technology of transistor packing hits a hard physical limit, certainly before we reach one molecule per transistor. However, it has already survived predictions of its demise. Each time some new technique comes to its rescue. Ultimately, components with periodic structures will be made by self-assembly, which means that the microscopic elements arrange themselves in an orderly structure much as molecules do during crystallization. If we give Moore's law a broad interpretation that encompasses derivative technologies, then this quasi-exponential trend could continue until it contributes to the end of our world.

All this progress has occurred on the two-dimensional surface of the silicon chip; the third dimension has yet to be exploited. Stacking integrated circuits vertically will create a serious cooling problem, but that is not a fundamental limitation; engineers will eventually solve it. We can reasonably expect another thousandfold increase during the next 30 years, only a little more than a human generation.

As the capability of digital hardware grows exponentially, one might think that the pace of artificial intelligence will be held in check by the time and intellect required for programmers to write the computer code. Not so. That process will also grow exponentially by means of genetic algorithms, a most amazing new technology. Computer programs exist today whose developers, the human ones, have no idea how they work. Because of limited education and intellect, humans may not be capable of knowing how they work! The actual software development happens inside a big computer equipped with a genetic algorithm (GA) that develops the program through an evolutionary process.

It begins with a seed program that does something vaguely like the desired program. Then the GA makes small random adjustments in the program's code and tests the result using known examples. (Suppose the problem is to factor enormous integers. GA would create test examples by multiplying big prime numbers to obtain integers to be factored.) Next, GA grades the altered program rather like a teacher grading a pupil. It keeps the programs with the highest scores and proceeds to the next generation with another set of adjustments. The process continues until a near-perfect program emerges. All of this evolution runs automatically at high speed inside the Central Processing Unit (CPU) of a supercomputer at rates trillions of times faster than biological evolution.

So far this description applies to asexual algorithms. Sexual ones involve two or more evolving programs that occasionally swap bits of code by analogy to biology. Distinctions between man and machine dwindle as fertilization *in silico* takes its place alongside fertilization *in vitro*. Robot manufacturers will compete using genetic algorithms to make their androids ever smarter. These algorithms will advance robotics software far beyond anything human programmers (or androids) can comprehend, much less compose as ordinary computer code [52].

#   #   #

When advanced robots arrive, people will quickly grow addicted to cheap, skilled slave labor. Pundits and politicians will urge caution and safeguards to no avail. They cannot restrain the demand for ever-smarter models. Some nations will outlaw manufacture or at least limit their permissible intelligence, but then another nation takes over the market and reaps the profits. Smuggling will be low risk because the robots can cross national borders on their own by walking, hitching rides, snorkeling across rivers, or whatever works. Their body shapes and camouflage coloring can be designed for self-smuggling. The juggernaut is unstoppable.

The serious threat is human hackers. They may deliberately breed a hostile strain of androids, which then infects normal ones with its virus. To do this, the hackers must obtain a genetic algorithm and pervert it, probably early in the robotic age before safeguards become sophisticated. And their efforts may be sponsored by a billionaire who thinks he hears orders from God.

Excluding hackers, it seems unlikely that androids will turn against us as they do in some movies. The engineers who develop them will be so concerned about hostility that they will build in safeguards at every level of their behavioral programs, in effect giving the robots powerful instincts to nurture humans. It is unlikely that the androids will violate such numerous and basic instincts. They cannot turn hostile in a single accidental mutation. Computer code for hostility is too complex; that would be like the proverbial monkey accidentally typing Shakespeare's works.

A series of mutations could gradually undermine the original safeguards and produce hostility, but only if some compelling Darwinian advantage reinforces the trend. Fortunately there is none. Androids do not eat human flesh nor compete for cropland or scarce resources. All the gentle androids need is minerals to make body parts and plenty of sunshine for their solar batteries. As for lebensraum they can design themselves to be content in the desert, on platforms at sea, or in the Arctic.

Nobody truly understands consciousness. The one who comes closest may be neuroscientist Gerald Edelman, Nobel laureate. He said, "Someday scientists will make a conscious artifact" [53]. He and mathematician Eugene Izhikevitch have already made a numerical simulation of a brain, which, to their delight, exhibits intrinsic activity. In other words, it is thinking on its own even when it has no assigned task or sensory input.

In the very long term, androids will become conscious for the same reasons humans did, whatever those reasons may be. Lazy humans may let them run their own genetic algorithms, at which point they control their own destiny and become a species practicing unlimited, unregulated, open-ended evolution. In the words of Richard Posner, "Human beings may turn out to be the twenty-first century's chimpanzees" [47].

Bill Joy fears the technology he helped create. Maybe he's right, but to me it seems more likely that robots will be our salvation. The same programmers who create safeguards against hostility are unlikely to protect human rights such as freedom, dignity, self-determination, justice, equality, and the unrestricted right to reproduce. The programmers simply do not think about these issues in the context of their daytime jobs. Even if they do, they will have great difficulty translating them into computer code. In summary, the androids have powerful instincts to nurture humans, but these instincts are unencumbered by concerns for human rights. Androids will feel free to impose a harsh discipline that saves us from ourselves while violating many of our so-called human rights.

World Simulations   A detailed unimaginative approach to survivability would make a huge numerical model of our entire world and run thousands of simulations of our future, each with slightly different inputs and random events. Statistics of the outcomes would then indicate the probabilities of survival at various levels of confidence. If world simulation seems far-fetched, then imagine living in 1975 and somebody predicts that in thirty years most middle-class homes will have an appliance that searches most of the world's knowledge in less than a second.

If current trends continue, computer capability will have increased many thousandfold in fifty years, which may be enough to enable the simulation. It will require at least three major modules: the physical Earth, human behavior, and the economy.

Existing computer programs simulate huge physical systems, for example world climate models. One component is a representation of oceans, a three-dimensional grid that specifies temperature, pressure, current, salinity, carbon dioxide concentration and contaminants within each volumetric cell. Similar representations describe the land and atmosphere.

Numerical models of the world economy date back to 1970 when the Club of Rome made a big simulation that included natural resources, population, money, industrial output, consumption, birth rates, food, environmental pollution, and much more. They called their report *The Limits to Growth* [54].

The big missing piece is simulation of human behavior to the level of individual personalities. Complexity theory tells us that tiny differences in the initial conditions (perhaps small childhood events) cause a huge divergence in the outcome. Therefore, any single run (execution) of the program will be virtually meaningless. However, the simulation programs will make hundreds or thousands of runs with slightly different initial conditions from one to the next. Then the full set of outputs will give the investigators a sense of the range of plausible responses they can expect from each simulated individual.

With this tool a diplomat can prepare for negotiations with her foreign counterpart by first negotiating with his avatar. After the real meeting she can revise the virtual personality and rehearse for their next encounter. Likewise, a candidate for public office can simulate a public debate with her opponent to prepare for the real one.

A world simulation cannot possibly treat eight billion people as individuals. However, it can lump ordinary people into groups with generic personalities: the membership of each labor union, immigrants from Mexico, the Caltech physics faculty, Hutu farmers in Rwanda, and so forth. But world leaders, corporate leaders, renowned scientists, and other influential and extraordinary individuals each merit their own unique virtual personality and behavior pattern. When the world model runs a short-term forecast, it will use personalities of named individuals. However, long-term forecasts must assume personalities for unknown future leaders. For any single run each personality is drawn from a statistical ensemble that covers the range of traits found among leaders of that nation, organization, or ethnic group.

Again the single run means very little. What counts is the range of possibilities established by a very large number of runs. Using a so-called Monte Carlo algorithm, the world model will run thousands of simulations in which parameters are varied within their range of uncertainty, especially the random personalities of future leaders. This builds a statistical database of plausible futures from which events can be predicted with varying degrees of confidence.

To be realistic, the world model must have imagination and "think outside the box". It must occasionally inject imponderables such as tipping points, paradigm shifts, and the unexpected geniuses and inventions that cause them. For example, in August 2001, no list of terrorist weapons included box cutters. Hence, a simulation could not predict the terrorist attack on September 11, but it might yield comparable disasters that indicate a level of risk and general areas for concern.

However, imagination can go too far by injecting too many extraordinary events. Certain parameters must be adjusted to realistic values: the frequency of tipping points, paradigm shifts, and revolutionary inventions, the percent variation in initial conditions, and similar quantities. This is a significant obstacle.

A possible solution would calibrate these parameters against historical data. Pretend to live in an earlier time and then run the program to "predict" events similar to those that have actually happened. This method is limited because history does not include extinctions and near-extinctions among the test cases. Moreover, it is doubtful that extant histoical records provide enough details to initialize the simulation.

A more practical approach would extract summary data from the simulations and compare them to an analytic formula such as our Equation 16. It is not likely that a seriously flawed simulation would conform to the formula. This test is analogous to Benford's law discussed in Section 1.5. It gives the statistics of the leading digits in a big set of numbers that are measurements of something. Benford's law has trapped embezzlers and tax evaders who have cooked their books with fictitious numbers that do not obey the law. In a similar manner an appropriate analytic formula may expose invalid results from a set of Monte Carlo simulations.

No simulation can replace analytic models for three reasons. The first is time; we cannot afford to wait fifty years for the model to operate reliably. Our species or civilization could collapse in the meantime. The second is complexity; it is hard for independent evaluators to digest all the inputs and intricacies and then make a valid assessment. The third is the calibration discussed above. The analytic model and the simulation are complementary. The latter provides the details with many surprising and important trends. The former provides the reality check. No matter how sophisticated simulations become, we'd better keep analytical models of survivability as reality checks. No doubt there will be successors to my model that will be more detailed and accurate but still within the analytic genre.

When the world model and the analytic model work together, this pair stands a good chance of producing valid forewarnings. Whether the public pays attention is anybody's guess. Scientists running the model would use their results to make a candid list of urgent reforms. It may include harsh measures that offend almost everybody: restrain the economy, levy heavy taxes on fossil fuels and other natural resources, impose compulsory birth control, and the like. The public reaction to such a report would itself be an interesting subject for a simulation.

One can imagine the repercussions: Special interests hire scientists willing to downplay or ridicule the world model. Celebrities like the late Dr. Michael Crichton will declare themselves instant experts and enumerate all the prior predictions of doom that have failed. They will find lots of them since each simulated future is merely a possibility, not a certainty. The real scientists will be branded as alarmists trying to inflate their own importance. The public is helpless because the simulation's flowchart alone is too complex for any individual to grasp. Few people have time to read it, much less verify the algorithms and statistical inputs.

Bulky data offer opportunities to insinuate biases consciously or not. An old mainstay, selection bias, emphasizes one class of data and downplays or ignores another. Potentially friendly critics know this game and remain skeptical. Most of them put the problem aside for another day, which somehow never comes. By contrast, hostile critics are motivated to sift through the data and inevitably find some that are suspect and a few inputs that have been questioned or discredited. (Think of creationists attacking Darwinism.) In the end the simulations will have little impact on our way of life regardless of their true potential. Still, we must try.

So, will the simulation's existence enhance or jeopardize our survival? It could go either way. After a simulation averts a few ordinary disasters, people may have a false sense of security. We press on with an expanding economy and ever-higher technology confident that our computer program will warn us when we begin living too close to the edge. Then the fatal one hits, something neither the machine nor its human supervisors had ever imagined.

5.3 OVERRATED NATURAL HAZARDS

Our sun behaved alarmingly from 1645 to 1715, the so-called Maunder Minimum when sunspots almost disappeared. This occurred during the four centuries of the Little Ice Age in Earth's Northern Hemisphere, which suggests a common cause. However, the sun recovered and displayed normal spots for the next three centuries, so our solar rotisserie seems secure for centuries to come. We now know more solar and stellar physics and see no evidence that the sun will misbehave, nor will a nearby star explode during the centuries in question. As they searched for sunspots, one wonders whether Edward Walter Maunder and his colleagues felt much anxiety for the world's future, especially since they lived during the Little Ice Age.

Volcanism can spew enough ash into the stratosphere to shade Earth and cause widespread famine. This happened in April 1815 when Mount Tambora in Indonesia erupted. It canceled the summer of 1816 in the Northern Hemisphere causing hardship in Europe and China and famine in New England [55]. Curiously the ash had little effect during the summer of 1815, which began two months after the eruption. Nor did it cancel summer in the Southern Hemisphere even though the volcano is situated at 8° south latitude. This was probably because the huge southern oceans comprise a heat reservoir like none other on Earth. (Incidentally, the so-called Year without a Summer spoiled Mary Shelley's vacation in Switzerland, so she passed the time by writing her novel *Frankenstein*.)

Geologists find evidence of catastrophic eruptions of far greater size that occurred at intervals of a few hundred thousand years. On that time scale our species will have likely expired from another cause.

Our survival for 2,000 centuries is a consequence of adaptation to killer plagues, famine, and other natural hazards of ancient times. By contrast, our exposure to man-made risks began only about a half century ago, which is shorter by a factor of 4,000. Based on this contrast Chapter 4 shows mathematically that natural hazards are insignificant compared to man-made hazards. But few people are familiar with this analysis, and even for those who are, the abstract argument lacks impact and drama. People are understandably in awe of nature's power—hurricanes, tsunamis, droughts, fire, and so on—and so they assume that nature is at least as threatening as humankind.

Concern about bolide strikes is almost a fad, especially since the comet *Shoe-maker–Levy* collided with Jupiter, thus drawing attention to the possibility. A loose international effort called the *Spaceguard Foundation* detects and tracks asteroids and other threatening objects that might possibly collide with Earth. Effective defense requires years of warning and careful planning. Objects with diameters bigger than a few kilometers will be tracked decades in advance of a possible collision. The threshold for extinction size is about 10 km in diameter, the probable size of the dinosaur killer. A lesser bolide might surprise us, but at worst it could depopulate only a modest-sized continent.

Many hazardous objects follow very predictable orbits and can be deflected harmlessly by any one of several techniques. A review in *Scientific American* describes half a dozen defenses [56]. For example, smashing a big spacecraft into the object at about the midpoint of its trajectory would nudge it into a harmless orbit. Exploding a nuclear bomb has been considered. In some cases a sustained gentle push or pull would suffice. If we dust the object with white pigment, it increases solar radiation pressure, which may provide enough gentle push. A feasible pull might be gravitational attraction to a nearby spacecraft of appropriate mass. This last scheme is totally independent of the physical characteristics of the object, such as its composition and spin. Thus the gravity scheme might be important in cases where we know nothing about the object's mechanical characteristics.

The most deadly objects, some tens of kilometers in diameter, are simply too massive to nudge. Fortunately they are also extremely rare, hence not in the same league with man-made hazards.

One exceptional hazard is a comet passing very close to the sun. Extreme heat boils gases off the comet's surface, which propels it slightly in the manner of a rocket jet. We have no way to accurately estimate the jet's thrust because the comet's shape, chemistry and physics are largely unknown. Thus, as it emerges from behind the sun, its orbit is unpredictable to the accuracy required to forestall a collision. While the comet is close to the sun, it is difficult to observe from the ground. Eventually we track it and compute its new trajectory, but by then there is little time to react.

However, all these bolide scenarios are details. The bottom line is that humanity has an extremely long record of survival in the presence of these hazards. Hence,

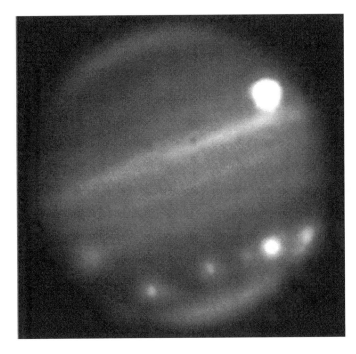

Impacts of Comet Shoemaker–Levy 9 on Jupiter. The bright disk in the northwest [*sic*] is the Jovian moon Io. The row of bright spots in the south is impacts of fragments of the comet. The dim spot in the southeast is the Great Red Spot (infrared image

bolides are overrated compared to hazards we have survived for merely half a century or less.

5.4 TRIAGE

Medical aid is usually inadequate in the aftermath of a great disaster or battle. To allocate scarce resources to best advantage, medics practice triage, which classifies the injured into three groups. Treatment is withheld from those who will survive without it, and also from those who will die regardless. Only the third group gets full attention, the ones who require treatment to survive.

There is so much misery in the world today that triage should be the guideline for philanthropy, but alas it seldom is. Consider, for example, the battle against AIDS in underdeveloped countries. If many cases can be prevented for the price of treating one, then prevention should get priority, while those already afflicted should be regarded as dead soldiers.

Discussion of prevention often arouses emotional distractions because it involves condoms. Machismo is one impediment. Another is the accusation by Third World leaders that western philanthropy has a hidden agenda to depopulate their countries. Let us set these issues aside and focus on triage purely as cost effectiveness of treatment versus prevention.

Recent price reductions in underdeveloped countries have brought the price of treatment per person down to about a dollar a day; however, this is still very expensive for anyone living on two dollars per day. Treatment must be perpetual because there is presently no cure for AIDS. As for prevention, a comparable effort to hold the price down will provide protection for about ten cents per day, one tenth the price of treatment. This makes an ideal case for triage: those already infected should be ignored, except for pregnant women with AIDS, who should be treated to save the baby.

A recent television report showed two ex-presidents, Bill Clinton and his good friend George Bush, Sr., endorsing an AIDS treatment program. They appeared with a stricken child in a Third World country. She was one of the few getting medical treatment as a result of the program. It was a heartwarming photo-op, the kind politicians love. Well, they are ex-politicians now, but old habits die hard.

The camera could not show a scene in the future where ten AIDS victims lie dying, those who could have been saved by spending the same money for prevention. Nor could the camera show the likely death of that same child at a future time when the drug delivery system breaks down in her village, a likely occurrence. Meanwhile, her presidential benefactors have moved on to other projects and will never know of her fate.

Surely Bush and Clinton understand the triage equation. So what was going on? Well, medical aid is held in high esteem, much higher than condom distribution. Moreover, some of the better places for real progress are in sordid waterfront bars and truck stops where prostitutes hang out—not great photo-ops for politicians. President Bush, Jr. maintained the same inept policy by spending $15 billion on AIDs relief, mostly in Africa, through his Pepfar program (President's Emergency Plan for AIDS Relief).

#   #   #

The issues in this treatise add a new dimension to philanthropic choices as shown in the Venn diagram, Figure 28. Philanthropy typically ignores the set on the left and scatters donations throughout the set on the right. In view of threats to the human race, it would make sense to concentrate most aid in the intersection.

An altruist might decide sadly but reasonably to abandon lifesaving efforts in overpopulated countries, because the long-term benefit is marginal. The people saved (and their future children) occupy space and consume scarce resources that others need, many of whom will die as a consequence. The net benefit of lifesaving may be positive; certainly those saved and their families appreciate their benefactors, trust them, and become influenced by western ways. However, the net gain may be too marginal compared to worthy causes in the intersection of the two sets, for example family planning in overpopulated countries.

Two lists of the top philanthropists [57] tell what causes they support. It is heartwarming to read about billions of dollars going to scholarships, poverty prevention, hospitals, schools, museums, and conservation. It was nice of Denny Sanford to pledge $5 million to the Crazy Horse Memorial Foundation, which is carving

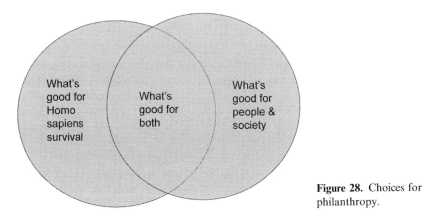

Figure 28. Choices for philanthropy.

a mountain to make a giant sculpture of the famous Lakota Sioux. But very few philanthropists seem aware that humankind is at risk, and that they can do something to help. An exception is Ted Turner, who funds projects for the environment and population and the Nuclear Threat Initiative. Also, the Bill and Melinda Gates Foundation gave $8.8 million to an organization that in my opinion does the absolute most good per dollar in the intersection of sets in the Venn diagram (Figure 28). And we must remember Nobel laureate Albert Gore for his dedication to global warming, even though he is not among the top fifty philanthropists.

5.5   REFLECTIONS ON THIS STUDY

Most studies that address the big questions of economics and ecology have no simple formulation like Gott's survival predictor, Equation 2 in Section 1.4, and so analysts resort to huge numerical simulations with myriad judgment calls. A classic example is the Club of Rome's study of *Limits to Growth* [54, 58]. A drawback of big models is that analysts (like most of us) are often partial to some outcome. Consciously or not they may bias their assumptions to produce a desired result. After publication, another analyst comes along with the opposite predilection. Her model and assumptions reach the opposite conclusion. The matter degenerates into conflicting schools of thought with no clear resolution, while the bewildered public has no time to delve into the bowels of the models and challenge the myriad details.

The results of this treatise support my proclivity for restraint as opposed to unfettered growth and development. However, unlike big numerical models, the predictor—Equation 16—is so simple that it gives opponents only two places to attack, and they are out in plain sight for all to judge. You can keep your daytime job and still find time enough to challenge this formula. The first vulnerable spot is the choice of cum-risk, $Z = M^{(\omega+1)} = (X - X_0)^{(\omega+1)}$, essentially the choice of parameters X_0 and ω. The equation for the current (2009) risk rate Λ, which appears in Section 4.6, is proportional to $(\omega + 1)/M_p$. My own attempts to evaluate this ratio by various means span no more than 15% change, but this range follows from my own mindset. Your mindset might bring the variation up to 30%.

The second uncertain quantity is $W(q)$, the relative weights of natural and man-made hazards, which is plotted in Figure 23. Details appear in Appendix L. A revised $W(q)$ can lean in either direction: a shift toward natural hazards (small q) is optimistic for human survival; a shift toward man-made hazards (big q) is pessimistic. In Appendix L (Equations L-3 and L-5) we use a very natural expression for $W(q)$, which involves a square root. When we change it to a cube root, which is somewhat contrived (L-4 and L-6), it increases the current hazard rate by 33% for species survivability, but only 7% for civilization. This change appears as dashed curves in Figures 23, 24, and 25. Combining the two vulnerable spots in the analysis might indicate errors as great as 50% in some of our results.

#   #   #

Opponents of my analysis will include optimists like the late Julian Simon [4], also the late Michael Crichton [5, 60]. People ensconced in comfortable ruts resist ideas that would require change. Dr. Crichton reviewed various predictions of doom that have failed, and suggested that all such predictions are invalid. He thought "inside the box", which contains a finite list of known hazards and the means for coping with them. Outside "the box" all sorts of unforeseen threats are lurking in our chaotic world. For example, the weapons of September 11, 2001 were box cutters and hijacked aircraft. Our complex biosphere has many possibilities for runaway positive feedback, some of which have never been identified. New hazards bombard us with increasing frequency until we eventually fail to anticipate the crucial combination in time to adapt.

The new anxieties—climate change, population explosion, exhaustion of raw materials, robotics, and pollution—arise from our evolved instincts that have successfully kept humankind going for 2,000 centuries. Crichton should not have disparaged them. He used his novels to express prophetic opinions. He failed in 1992 [59] by warning America about economic domination by predatory Japan. Now he has failed again with a novel [60] that portrays environmentalists as the bad guys. People should read his novels for entertainment, nothing more.

5.6 PROSPECTS FOR A SAFER WORLD

Sir Martin Rees [2] has wagered $1,000 that "by the year 2020 a single instance of bio-error or bio-terror will have killed a million people." He writes that he fervently hopes to lose that bet. Humane and politically correct people are supposed to share that sentiment, and yet, as we saw in Chapter 4, a near-extinction event is required to save our species. Is there no humane way out of this dilemma?

Population reduction would solve everything. Census extrapolation for world population peaks at nine or ten billion in about 2090. From there the decline may continue, perhaps to 4 billion (the world population in 1974), or it may falter and let the population creep up to 12 billion. The latter seems more likely. Optimists say that the big population surge is ending; the long-awaited "demographic transition" is catching on [61]. True in the near future, but this is simplistic extrapolation with no appreciation for *tipping points*.

In the long run, fertility decline is probably just a hiatus. We're seeing the result of the contraceptive revolution. By enabling sex without pregnancy, it defeats nature's strongest reproductive strategy. But nature has other tricks in reserve. They will not be apparent for a while, because it takes several generations of natural selection to make them conspicuous.

A friend of mine planned to have three children, but the third was a set of triplets. His two girls and three boys comprised a *full house*! His many grandchildren will propagate the predisposition for multiple births, especially multiples in the second or third pregnancy. Twins from the first pregnancy do little to inflate family size because the parents simply quit at that point. But when multiple births happen later, they do inflate family size as happened to my friend.

Other pronatalistic incentives include provision for old age, and a simple love of children—lots of them. Natural selection will reinforce nature's backup tricks until one of them restores the population explosion.

Any one pronatalist group acting alone can overpopulate the entire world. Most of these traditional societies and religions will mellow out and join the mainstream. Catholic Italy and Spain have birthrates well below replacement. Even the communal Hutterites have dropped their fertility from about ten to about five or six, which is still alarming [62]. But to permanently stabilize population, every last one of these groups must change, and it is hard to believe this could happen voluntarily.

The United States and Canada have unwise population policies. They admit almost anybody—many claiming to be refugees—who then reproduce freely. People like me are to blame. Both my housekeeper and my gardener are immigrants, each with four children. I should fire them, but I like them personally—their children too. It is one thing to be an armchair pundit, quite another to practice what you preach. Multiply my personal fault by a billion or two, and you have the makings of extinction.

Optimists have faith in benign social programs: education, contraception, and incentives for sterilization. These work for a while, but in the last analysis they are just a form of selective breeding. The surest way to breed stubborn pronatalists is to dissuade everybody else from reproducing [63]. As cooperators die off, they vacate land and resources, which the pronatalists gladly take for their big families. Humanists seem to ignore basic rate equations for exponential processes that engineers use routinely.

#   #   #

Somewhere in our galaxy a few humanoid species may have broken out of their home planets and expanded to other habitats or other stars. These winners must have made a concerted effort. Perhaps narrow escapes from extinction have focused them on a goal and spurred them to action, or perhaps their world governments imposed the goal by force.

We earthlings dislike coercive world government; our leading nations and the U.N. are committed to reproductive freedom and self-determination, which I am sure most of my readers also support. But this benign ethic was forged by goodhearted people with the best of motives who never saw Figure 27. Let Gott's predictor tell us about the survivability of this ethic. The start date was fuzzy; let us say 1960. Before that time, colonialism, racism, and various other forms of suppression were common-place. So let us put age 48 years into Equations 3 and find,

duration of benign ethic > 5 years with 90% confidence;

> 48 years with 50%.

The next ethic may be almost anything: good or evil, safe or risky. Hopefully a population homogenized by intermarriage will reduce tensions among groups. How-ever, we may get imperialism redux, ethnic cleansing, subsidized joy pills doped with contraceptives, harsh state-imposed limits on family size as in China, or reproductive quotas for various ethnic groups. Nobody knows.

#   #   #

We tend to rank perils based on emotion rather than reason. In the United States we celebrate Halloween (October 31) by letting 36 million children put on costumes and extort candy from neighbors, including strangers. A few decades ago there was a big scare that people were putting needles, pins, razor blades, and sometimes poison in the candy. Many parents forbade their children to eat anything that was not a commercial brand in its original wrapper with no sign of tampering. The poison stories all turned out to be hoaxes. There were only about a dozen instances of very minor injury from sharp objects.

Many of these same parents take their children in the car while driving on routine errands, a far greater risk. Moreover, these children are subject to greater risk simply by being earthlings in the 21st century. A cataclysm that wipes out civilization will kill vast numbers of them, and in Chapter 4 we found that hazard rate to be about 1% per year.

During humanity's 2,000 centuries, people evolved a strong concern for survival of their own tribes, maybe even nations, but nothing as vast as their whole species. Now, for lack of such an instinct, there is no sense of urgency or political will for a trillion-dollar international project to save humanity.

I am no different. During a productive phase of this survivability project, I was offered a lucrative consulting job. Of course, I took it. Humanity be damned, here was a chance to make money! It is the same old Darwinian priority: first take care of self and family. Abstract thinking about the world at large has low priority because it has never before been a survival issue. It is a mere intellectual exercise rather than a powerful instinct.

Saving one person brings out powerful instincts in primitive parts of the brain. In prehistoric times the one saved was usually a fellow tribesman carrying many of the same genes as his rescuer. But today, saving humanity is a small intellectual idea somewhere among the brain's most recent updates. Perhaps Josef Stalin was thinking of this when he remarked, "A single death is a tragedy; a million deaths are a statistic" [64].

And so the short answer to the big question is, No. There is no way out of our dilemma. An apocalyptic event, perhaps a near-extinction, is prerequisite for long-term human survival, and that's just the way it is. Meanwhile, let's be jolly, enjoy some gallows humor, and let children accept candy from strangers.

5.7 SYSTEMIC STRENGTHS AND WEAKNESSES

The probability analysis in Chapter 4 applies to a hypothetical ensemble of humanoid species scattered about the galaxy, each member of which has passed through a phase similar to the one we earthlings are in now. Within this ensemble we have no idea what percentile of longevity we fall in. We differ from those other humanoids in ways that make us more or less survivable than typical members. Let us speculate on some of those variations.

This comparison will be too pessimistic because we cannot imagine the strange irrational thoughts and actions that may plague extraterrestrial humanoids. Like us they carry obsolete baggage from their prehistoric times including instincts that served their ancestors well but are superfluous or even a liability in a civilized environment. By contrast, it is much easier to identify our own irrational behavior. In this sense comparing earthlings to extraterrestrials resembles the comparison between ourselves and our adversaries (us versus them) in such events as war and business competition. We tacitly assume that our adversary is more rational, capable and knowledgeable than we ourselves, which usually proves to be untrue in the aftermath when all the facts come out.

Intellectual extremes One way to have a fairly safe society is to forbid all firearms. Another way is to require everybody to carry one. The middling condition is the most dangerous, which partially explains why the United States has so much crime. Similarly, an ignorant species is incapable of self-extinction. A genius species will protect itself from the hazards it creates. The most vulnerable species is the one with a broad spread of intellect, and this is what we have, especially in the United States.

Creative geniuses gave us nuclear power, lasers, fiber-optic communications, integrated circuits, computers, the Internet, search engines, genetic engineering, and robotics. These ingenious inventors usually manage their creations responsibly when they retain control of them. Some of them do become entrepreneurs and retain some measure of control of their inventions; Gordon Moore comes to mind. However, patents expire, competitors appear, and control slips away. Genius happily moves on to create more dangerous toys while control of the current ones devolves to business people and politicians, including those with inferior intellect and selfish motives. Then the cycle repeats. Perhaps other humanoid species have devised a better system of governance that avoids this hazard. If so, their futures may be more secure than ours.

Homogenization It is important that isolated tribes be left alone to live as they always have. Something in their diet, culture, surroundings, customs, or genetic code may give them immunity to the extinction hazard that will kill the rest of us. These tribespeople are humanity's backup population. Contact brings them drugs, new diet, microbes, and manufactured goods (cargo). One of those imports might somehow cancel their immunity and make them as vulnerable as the rest of us. Hence, it is vital that they be neither persecuted nor assimilated.

Misguided altruism Mosquitoes and other pests protect much of the remaining wetlands and jungle in the overpopulated Third World, but then well-meaning aid workers exterminate the six-legged defenders out of compassion for the burgeoning human population. In the battle between nature and humanity, we should favor nature, not for nature's sake but for humanity's sake. Nature often tries to restrain humanity and keep us more survivable, but then goodhearted people thwart her efforts for misguided short-term altruism.

Neoclassical economics A few economists are green [65, 66]. The late Kenneth Ewart Boulding was the most notable with metaphors of cowboy economy (reckless, exploitative) and a spaceman economy (constrained by the limited reservoirs of his spaceship) [67]. Unfortunately, most mainstream (neoclassical) economists are committed to growth and regard sustainability as a fad. Then there are globalization folks, who help the underdeveloped world to develop because this will create more markets for goods from the First World, as though we didn't already have enough.

Adaptation The event that killed the dinosaurs was something they experienced for the first time. Had there been many precursors that killed some fraction while others survived, then that bit of selective breeding would have enabled more species to survive the big hit. We might have descended from dinosaurs and might still exhibit vestiges of the reptilian line, perhaps eggs, scales, or something else.

As far as we know, our species has never in its 2,000 centuries been exposed to an apocalyptic event of that magnitude. That leaves us much more vulnerable than other extraterrestrials who have survived one and recorded it in their history. However, there is a curious twist: even without their historical motivation, we somehow developed an interest in the subject. What sort of evolutionary selection acted on our ancestors to make me want to write this treatise and you want to read it?

Curiosity is a factor. It had obvious survival benefit when our ancestors investigated strange footprints near the tribal campsite. Curiosity extended to scary things that proved harmless like eclipses and rainbows. It produced practical inventions with great survival value. Now we carry this trait to extremes by sending expensive spacecraft to observe not only earth-like bodies but also obscure ones that tell us nothing of any practical value. To my knowledge nobody has explained how we evolved such extreme curiosity. Yet it may save our species by causing us to investigate extinction scenarios that are completely beyond our experience and history.

Somewhere a humanoid species may have as much high technology as we do, but no instinctive drive to think beyond practical applications, no desire to ask questions about the big picture, things like species survival. Those types are more vulnerable than we are.

Solutions to social problems It would seem natural to find sustainable social solutions for social problems. Instead we evade social change by adapting hazardous technical solutions:

- Our government could create incentives for people to donate organs for transplant, but instead we breed special pigs to supply scarce organs, a procedure called xenotransplantation. Obviously this is an opportunity for xenogenetic diseases to jump from pig to man. Such a disease may be mild in pigs and fatal to humans, just as simian immunodeficiency syndrome (SIDS) is mild compared to AIDS.
- Cities could be made safe and livable, but instead people emigrate to suburbs and commute daily to the city thus wasting time and money and needlessly generating carbon dioxide.
- At rush hour our freeways are jammed with full-sized cars, the great majority of which carry only the driver. So we widen freeways and build more of them. A creative social solution would make hitchhiking safe, desirable, and a social obligation for those who own cars, just as tipping waiting staff after good service is a social obligation. For safety, an ID system could keep records of who traveled with whom. It is quite possible to manage these data in the age of computers and cell phones.

Mental disconnect I have given three lectures on human survivability to scholarly groups with varied specialties. After presenting my calculated results I noted that the survival risks are comparable to ordinary perils that insurance companies underwrite. All three audiences calmly accepted that and then asked good intellectual questions about my analysis. Ironically there is a whole class of questions that nobody asked—not one, not even close; questions with intense personal concern like the following:

- How can we initiate political action to make a safer world? Do you think any existing organization will take up the cause?
- What are the chances that death will be quick and relatively painless?
- My son and his bride plan to have children. Should I discourage them?

Perhaps my audiences didn't believe me and tuned out. I tested them with the joke slide in Figure 29. I tried to say, "Since survival risks are comparable to insured risks, somebody ought to sell extinction insurance. However, I'm the only person with actuarial tables, so I must take on that burden. Now if any of you brought your checkbooks ...". But milliseconds after the slide appeared, laughter drowned out those words. So they were attentive, and the question remains: Why no expression of intense personal concern? No doubt these same people would be troubled if they accidentally let an insurance policy expire.

Limited time offer—

EXTINCTION INSURANCE

Act now! Lock in this rock-
bottom introductory premium.

$1,000,000 policy now only
$995 per year.

Figure 29. Extinction
insurance.

My audiences reacted as though I deserved credit for solving an amusing mathematical puzzle. Perhaps it went straight to one cerebral hemisphere where intellect resides but never reached the powerful emotions and motivators in the other side. Apparently the topic trips a circuit breaker in the corpus callosum and shuts down the connection.

Between lectures I discovered a much better way to formulate the cum-risk Z discussed in Section 4.2. This made my analysis more accurate and convincing, so I was elated. Oops, I forgot to be sad because the risk is three times my previous estimate, which jeopardizes my own grandchildren! Quite clearly I have the same mental block as the folks in my audiences.

If other species in our galactic ensemble have no such mental block, then they may be more survivable than we.

5.8 SECOND CHANCE?

If a great calamity destroys civilization, the survivors will undoubtedly rebuild. No matter how primitive they may be, they will discover our artifacts, repair or replicate many of them, and proceed toward urbanization, industrial revolution and eventually science and technology.

But what about extinction? Will Earth get a second chance to evolve a humanoid species that develops civilization, industry, and scientific curiosity? Evolution would take many millions of years compared to only a few centuries to replace civilization. Could we be just such a replacement, the second rather than the first species to occupy the humanoid ecological niche? Not likely, paleontologists would surely have discovered ruins or some sort of evidence if a predecessor had existed.

Another fact guarantees our status as first. Our ancestors found *native metals*, almost pure stuff that is ready to use with little or no metallurgy. We have now exhausted all such handy resources and must obtain metals from ores, mostly sulfides and oxides. Had there been a prior humanoid species, they would have exhausted the native stuff, and our ancestors would have found none.

Native copper, 4 centimeters (courtesy of Jonathan Zander).

In particular, our ancestors found *native copper* and fashioned it into useful artifacts. Sometimes they found nuggets in creek beds or in pits in the ground, and sometimes irregular chunks of copper embedded in rock. On rare occasions they found metallic gold, silver, and platinum. Other metals, such as zinc, tin, and nickel, were scarce in the pure native form but abundant as ores. Native nonmetals, especially sulfur, have been found in abundance, also metalloids, which include arsenic, bismuth, and antimony. Our ancestors would have found nothing so convenient had they not been first.

So, what if anything will take over the humanoid niche after we expire? Perhaps some rodent will survive the extinction event and eventually develop hands, start walking upright, develop a big brain, and ultimately move into the humanoid niche. This evolution can proceed to the level of industry based on stone, wood, leather, and other abundant resources. They will invent the windmill, waterwheel, and such. But then their evolution hits an insurmountable barrier for lack of readily accessible resources such as native metals and rich ores. Without them there is nothing to spark their curiosity and give them a vision of a world with things more interesting than wood, leather, and stone.

#   #   #

Perhaps the first human coppersmith became wealthy, took two wives, and begot many children, some of whom inherited his inventive skills. At that point in the long complex Darwinian process, one strand turned a corner toward evolution of modern man.

Another ancient man discovered how to exploit oil he found leaking from cracks in the ground. Yet another became the first smelter. He built a fireplace from unusual rocks. During a particularly hot fire, molten metal oozed from the rocks where it came in contact with embers. In this reaction hot carbon reduced a sulfide or oxide ore leaving a puddle of metal, probably lead since other ores require higher temperature than a basic fireplace achieves.

Our replacement species will not have such educational experiences, nor will they find such handy resources. The nascent humanoids that replace us will be deprived of the stepping-stones they need to go on. With hardship they may be physically capable of finding the buried low-grade ores and deep pools of oil that we use today. However, no tribe living at a subsistence level would allocate the time and effort for such a project without a clear objective and a good prospect of success. Besides, at their state of development, they do not understand such concepts as research and long-term investment. An essential piece is missing, a teacher who could persuade them that the effort would be worthwhile.

Without a quick payoff, maybe one generation, it is hard to imagine how Darwinian selection could begin to reinforce the inventive skills required for an industrial revolution, much less hi-tech. Instead, competing needs would drive their evolution toward other benefits such as physical strength, warrior and hunter skills, and resistance to disease.

Perhaps another planet can produce more than one humanoid species if their resources for industry and hi-tech are as abundant as wood, leather and stone. But Earth can bear only one.

5.9 SURVIVAL HABITAT

Before our time expires, we may send colonists to a survival habitat. Instead of outer space, it is far cheaper to use an isolated or sheltered place on Earth and far easier to rescue anyone in trouble. But where? John Leslie [68] thinks we should build human survival colonies in artificial biospheres. He deplores the lack of any such project, and I agree.

However, the safest and most affordable habitat may be unfit for humans, but quite suitable for our sentient robotic successors. Their bodies can be designed especially for the new environment, either on post-human Earth or in outer space. Prof. Hans Moravec at Carnegie Mellon University would be comfortable with that. He regards our robotic successors as our Mind Children [69].

A colony on the sea floor next to a hydrothermal vent [70] might possibly become self-sufficient. (My Google inquiry turned up the current *Aquarius* program [71], but nothing that attempts to wean itself from the world above.) That secluded neighborhood should survive almost any calamity on the surface. Geothermal power is available for general needs and for making oxygen from seawater. A temperature differential from about 400°C in the vent to 2°C ambient is enough for a Carnot engine to run at 60% efficiency. Among the strange organisms attracted to the vents, some are probably edible, but to my knowledge, no one has yet done the taste test. Available power might support light industry but probably nothing heavy.

Besides formidable technical problems, there is a psychological obstacle to building a survival habitat, namely severe lack of urgency. Short-term concerns always take priority over those with a time scale of centuries—right up to the day disaster strikes. During 2,000 centuries of human life we have evolved a sense of urgency to deal with short-term hazards before they get out of hand: smoke in the distance; neighbors encroaching on tribal lands, and that sort of thing. But we have no such instinct for a threat that lurks in the background for a century or more. If this inadequacy is typical of humanoid species throughout the galaxy, it suggests an answer to Enrico Fermi's famous question: "Where are they?" They died because they were always too preoccupied to anticipate long-term consequences.

Once a colony escapes Earth's gravity at great expense, it makes little sense to trap it again in the gravity of another planet unless that planet has water or other resources more important than any we have found yet in the solar system. Hence, the best scheme for a space habitat may be O'Neill's cylinder [72]. The living space is the interior of a cylinder, which spins to make artificial gravity. Both gravity and climate are adjustable for comfort. The cylinder can begin life in Earth orbit where help is near. If and when it evolves into a permanent colony completely weaned from Earth, then the cylinder can migrate into solar orbit. Eventually the colony can mine the asteroid belt for material to expand or replicate more cylinders.

Sir Martin Rees [2] questions the feasibility of O'Neill's cylinder because it is vulnerable to sabotage, but that problem may be soluble. A saboteur needs privacy both to prepare his attack and to conceal his emotional state. Hence, privacy should be strictly limited to brief bodily functions (including sex). This is not an unnatural state. Primitive tribes had little privacy. Nor do sailors in the United States Navy, where officers inspect both workspaces and personal lockers. Violent incidents occur occasionally, especially with disgruntled drunken sailors, but rarely if ever has any-one threatened the ship. If anything could drive a mariner berserk, it would be claustrophobic life on a submarine, and yet carefully selected submariners adapt to patrols that last two months. (I spent one day submerged—the most I could possibly tolerate.)

NASA and other space agencies and space enthusiasts everywhere seem to over-look an important experiment: Establish self-sufficient colonies in harsh locations here on Earth. The idea is not to choose a site that simulates a particular destination in the solar system, but rather to develop intuition and experience in the problems of self-sufficiency. Hydrothermal vents mentioned above seem ideal. Other possibilities include Antarctica, the central Greenland ice sheet, and/or a harsh desert. Compared to space stations and planets, these all have huge advantages of familiar gravity, protection from space radiation, and breathable air at normal pressure. These colo-nies would provide a baseline for evaluating questions about launch weight (initial supplies), shelter, temperature control, food and other necessities, and in general, the difficulties of achieving self-sufficiency. *They also serve as survival colonies without calling them that.* Preparation for "space colonization" makes a much sexier pitch for funds than "survival colony", and the price is a bargain compared to other projects in the manned space program.

We should not wait too long to build a space habitat, because the window of opportunity will close (as Gott noted). Let us apply GSP to the duration of the space age. So far it has lasted (from Sputnik, 1957) $T_p = 52$ years, the same as Antarctic studies, hence,

$$\text{future of the space age} > 6 \text{ years with } 90\% \text{ confidence}$$

$$> 52 \text{ years with } 50\%$$

Space exploration may end with humans stranded on Earth just as the Rapa Nui (the people of Easter Island) were stranded when Captain Cook visited in 1774. Prior to that contact, Dutch admiral Roggeveen had found a hardy people in 1722. Some-time between those dates civil war and ecological collapse had apparently decimated both the population and the island's resources [73]. They had toppled and desecrated their stone statues, just punishment for gods who failed them. The trees were gone, also their seeds, and they had no material to build seaworthy vessels for escape.

Maybe Rapa Nui is a microcosm for humanity. If we spend all our resources on wars and extravagant living, then we may not have enough left to build the giant spaceships required to launch a viable colony. We may be marooned like the Rapa Nui. If this fate is normal for humanoids throughout the galaxy, it may resolve Fermi's paradox.

Richard Gott wants us to colonize Mars and then move on to planets throughout the galaxy [74]. He argues fervently that humans need a second planet (or other base) to insure survival of our species [75]. But he should know better! He famously chides people who think they are very special [76]. Yet humans would be very special indeed if we become the first earthly species out of millions to colonize another planet. Instead we should strive for a less special goal that is more attainable, to be the progenitors of the first species to colonize the galaxy—our robotic descendants, the androids.

We (or they) can design their bodies to be spaceworthy, able to thrive in a hard vacuum without water or oxygen. And they can be designed to hibernate throughout centuries of interstellar travel. It is far more feasible to design the colonists' body to thrive on a particular planet than it is to terraform the entire planet to accommodate our demanding maladapted bodies. There is a good chance that we will not be able to afford spaceships for humans but can afford them for androids. Look around any superstore and compare the number of things that humans need to the number that androids need. Then think about sending all that stuff to Mars for a few hundred thousand dollars per pound.

We may simply be living in the wrong times for big high-tech engineering projects. Manned spaceflight to the Moon and beyond has been on hold since the last Apollo mission in 1972. More than half the missions to Mars have failed. The space shuttle has been disappointing, and we have not built the proposed space elevator (lift). (This would employ a cable car riding up and down a cable that dangles from a space station above synchronous orbit to a point on Earth's surface.)

Our cities are not covered by geodesic domes that would be useful as skyhooks. Nuclear electric power is on hold. We have explored only tiny samples of the oceans' floors and the insides of mountains. Boreholes to Earth's mantle are challenging, and earthquakes still cannot be predicted. By contrast, the microelectronics and software needed to make android brains has been succeeding beyond our wildest dreams of a few decades ago. It seems clear where we should place our bets.

It would be a shame to populate a galaxy with our own pitiful, frail bodies, which require special care and feeding, major water supplies, crops, and so forth. Comfort and pleasure cost far more. Rather than gamble on a long shot, we should gracefully accept our limitations and take pride in our robotic descendants as they go forth to colonize the galaxy.

If we humans ever journey into the galaxy, we may travel as frozen embryos in the care of robo-nannies. This makes good ethical sense. If a mission fails, then the only human loss is a batch of embryos, not sentient adults who have endured a long journey only to perish disappointed. As for sentient androids on a failed mission, they can be reinsilicated here on Earth using backup copies of their brains and memories. When a mission succeeds, the androids will go out in the poisonous air and do all the building and exploring while their human pets live in a comfortable zoo with climate control. It will be a "brave new world" with little role for human courage and prowess.

Appendix A

Survival formula derived from hazard rates

Here we use probabilities of unknown hazard rates to derive formulas for survival. Section 1.1 summarizes the concept and results. First consider the case in which the hazard rate λ, probability of expiring per unit time, is fixed. Radioisotopes are a well-known example. The fraction of individuals in a sample that survive at time t (or the probability of any one surviving) is the well-known exponential decay:

$$Q(t \mid \lambda) = e^{-\lambda \cdot t} = \exp(-\lambda t) \qquad \text{(A-1)}$$

This formula applies to any entity if the hazard rate is constant in time: the entity neither learns how to survive, nor does it wear out.

Now suppose we do not know λ. For an atom this means that we do not know which of several discrete isotopes it might be; see Section 1.1. But for a stage production the possible λ's are continuous, and we estimate their probability density to be a function $F(\lambda)$. (This is like the abundance of isotopes in Figure 3.) Now the joint probability density that the system's intrinsic risk is λ, *and* that the system survives that risk for time t is $F(\lambda)$ *times* $\exp(-\lambda t)$. In survival problems we are typically indifferent to λ: we just want to know how long the entity in question survives. In probability theory you sum or integrate over random variables to which you are indifferent. (If you roll a die, the probability of getting a 5 OR a 6 is $1/6 + 1/6 = 1/3$.) Hence, the probability of survival for time t regardless of λ is

$$Q(t) = \int_0^\infty F(\lambda) \, e^{-\lambda t} \, d\lambda \qquad \text{(A-2)}$$

which is analogous to the mixture in Figure 3, the bold continuous curve.

We want an expression for $F(\lambda)$ that represents complete ignorance of λ. The most natural probability density is a uniform distribution; however, at this point we cannot rule out certain other functions. A minor problem is that when F is some finite constant, then $\int F \, d\lambda = \infty$. In other words F cannot be normalized to 1.0 nor any finite value.

To work around this problem, define the uniform distribution as

$$F(\lambda) = \mathrm{Lim}_{\tau \to 0}(\tau \cdot e^{-\tau \cdot \lambda}) \quad (\text{as } \tau \to 0, \exp(-\tau\lambda) \to 1.0) \tag{A-3}$$

Now

$$\int_0^\infty F(\lambda)\, d\lambda = 1 \quad \text{as required.}$$

If we delay putting $\tau = 0$ until later, Section 1.4, it will cause no problem.

However, there is a rationale for keeping $\tau > 0$. We cannot define the birth of a stage production within one second whether it be signing a contract or curtain rise on opening night, and so it makes no sense to allow $\lambda > 1/1$ sec. For every entity there is some maximum hazard rate that is plausible. A stage production does not fail while the curtain is rising on opening night, unless an explosion wrecks the theater at that instant. Likewise, a business does not fail while its owner is unlocking the door to admit her first customers. We cannot define a sharp maximum cutoff for $F(\lambda)$, so let it be gradual. Let τ denote a time at which the risk has dropped to less than half, perhaps several hours. Then removing the limit from Equation A-3 leaves

$$F(\lambda) = \tau\, e^{-\lambda \tau}; \qquad \int_0^\infty F(\lambda)\, d\lambda = 1 \tag{A-4}$$

Using Equation A-4 in A-2 gives the prior probability Q:

$$Q(t\,|\,\tau) = \frac{\tau}{t + \tau} \tag{A-5}$$

Note that $Q(0\,|\,\tau) = 1.0$ denoting 100% initial survivability in accord with the definition of Q.

The question of taking the limit $\tau \to 0$ hinges on the objective, namely finding a survivability formula for an entity about which we know nothing. If we're being strict about "nothing", then we must take the limit; otherwise, we know something, specifically an estimate of τ. But if we relax that stipulation and allow a bit of common-sense knowledge, then a finite estimate of τ improves the accuracy simply because we're using what we know rather than deliberately ignoring it.

However, many problems (including human survivability) do not involve the entity's infancy, and so for brevity people often neglect τ and simply use $Q = 1/t$, as prior probability, secure in the knowledge that posterior probability will get rid of the singularity and take care of normalization; compare Equation B-3.

When we keep $\tau > 0$, its definition above Equation A-4 matches that of J, the gestation period, so-called because a fully descriptive name would be too long. The connection to gestation is that the fact of birth has already eliminated hazards that would have caused miscarriage had the entity been susceptible; however, nothing prevents J from being somewhat greater than gestation requires.

In what follows we have no further need for τ, so let us put $\tau = J$ and proceed.

#   #   #

Why should $F(\lambda)$ be (almost) uniform? Why not some other probability distribution? Consider for example,

$$F = \sigma[1 + (\sigma\lambda)]^{-2}$$

The parameter σ, having units of time, is either a property of the entity or a universal constant. If it is the former, then its value is knowledge we are not supposed to have. And there is no appropriate universal constant. Decay times vary from femtoseconds for an excited atom to millions of years for a mountain range, so no universal σ could possibly apply over that range.

Besides this example, any other function will have this same problem except for a power law λ^p, the so-called scale-free case. (In the example above σ would be a time scale.) Normalizing this power law and putting $\tau = J$ gives

$$F(\lambda,p) = \frac{J^{1+p}}{\Gamma(1+p)} e^{-J\lambda}\lambda^p; \quad p > -1 \tag{A-6}$$

When $p = 0$, this is again the (almost) uniform distribution, Equation A-4. Using Equation A-6 in Equation A-2 gives

$$Q(T,p) = \left(\frac{J}{J+T}\right)^{1+p}; \quad p \geq 0 \text{ (see below)} \tag{A-7}$$

In the case $p = 0$ this is Equation 1.

Theoretically p is allowed in the range $-1 < p < 0$, but for a practical reason, this cannot be. Appendix C shows that for $p = 0$, the mean survival time is *soft infinity*, which means that the small aging effects we have neglected will ensure that the actual mean is finite. But $p < 0$ would make the mean a *hard infinity*. Then we would be seeing stage productions from at least the tenth century, and our world would suffocate under an accumulation of ancient stuff that doesn't expire fast enough. And billion-year-old exohumanoids (if any ever existed) would have invaded Earth for lebensraum.

According to the approach taken here and in Section 1.1, a small positive value of p is quite permissible. Since the factor λ^p vanishes at $\lambda = 0$, Equation A-5 says that risk cannot be exactly zero. In other words, nothing lives forever, which is certainly true. However, a different approach in Section 1.5 requires $p = 0$ as we assumed at the outset, Equation A-3. Conversely, Section 1.5 has a different weakness that this approach resolves in the following section.

A.1   VARIABLE HAZARD RATE

So far the hazard rate λ has been constant, which leads to Equation A-1, but this is merely a familiar example. Next we show that our predictor is more general and applies to cases in which hazard rate $r(t)$ is not constant. Now when r is known, what replaces Equation A-1 is

$$Q(t|r) = \exp(-Z(t)) \tag{A-8}$$

where Z is the cum-risk:

$$Z(t) = \int_0^t r(s)\, ds \qquad (A\text{-}9)$$

(This is well known and also easy to prove as follows: In the nth small time increment, Δ, the probability of expiring is $r_n\Delta$ and of surviving is $[1 - r_n\Delta]$. The probability of surviving all increments is the product:

$$(1 - z_1\Delta)\cdots(1 - z_n\Delta)\cdots = \exp(-z_1\Delta - z_2\Delta\cdots)$$

This becomes Equation A-8 in the limit $\Delta \to 0$.)

Next, replace $Z(t)$ by $\lambda \cdot Z(t)$, where Z denotes a known time dependence of the cum-risk and λ an unknown magnitude of the risk. Then Equations A-8 and A-2 become

$$Q(t\,|\,z) = \exp(-\lambda \cdot Z(t)) \qquad (A\text{-}10)$$

and

$$Q(t) = \int_0^\infty F(\lambda)\, e^{-\lambda Z(t)}\, d\lambda \qquad (A\text{-}11)$$

Integrating as before (with $\tau \to J$) gives

$$Q = \frac{J}{J + Z} = \frac{1}{1 + Z/J} \qquad (A\text{-}12)$$

which is the equation in Section 1.3 that we set out to prove.

A.2 UNKNOWN VULNERABILITY

As a final example consider the case in which the unknown variable is the vulnerability of the entity to a fixed risk.

Let $Y(t)$ denote a pre-prior survivability that applies to a typical individual entity in some class or species of interest. If Y has a tail longer than $\exp(-\lambda t)$, Equation A-1, it means that risk decreases with age as though the entity learned survival skills during adolescence. If the tail is shorter, the entity wears out to some extent.

Among the individuals in a statistical sample, some are frail and others are hardy, which we express by using a multiplier μ to stretch or compress their lifetime. Survivability of the frail ones is $Y(\mu t)$ with $\mu > 1$, which says that they age quickly. Likewise survivability of the hardy ones is $Y(\mu t)$ with $\mu < 1$. Our observer, who notes only the ages of individuals, cannot tell the difference. Besides, we assume that the observer knows nothing about the species, and so he has no idea how fast individuals age, except what he can infer from observed age. Hence, he takes an average as in Equation A-2:

$$Q(t) = \int_0^\infty F(\mu)\, Y(\mu t)\, d\mu \qquad (A\text{-}13)$$

We use almost uniform distribution again, Equation A-3, but with μ instead of λ:

$$Q(t\,|\,J) = J \int_0^\infty Y(\mu t)\, e^{-\mu J}\, d\mu = \frac{J}{t} \int_0^\infty Y(z)\, \exp(-Jz/t)\, dz \qquad \text{(A-14)}$$

If the integral,

$$I = \int_0^\infty Y(z)\, dz \qquad \text{(A-15)}$$

converges to a finite value, then as before, J/t in the exponential factor, Equation A-14, is so small at birth and afterward, that we are justified in calling it zero, in which case Equation A-14 becomes

$$Q = IJ/t \qquad \text{(A-16)}$$

Interpreting t as time from conception $T + J$ again leads to Equation 1 as we set out to show.

However, if the integral in A-15 does not converge, we must find a finite estimate of J and use it in Equation A-14. Fortunately this does not happen in practical cases as an interpretation of Equation A-15 will show. Let us integrate by parts:

$$I = \int_0^\infty \{-Y'(t)\}t\, dt + [t\,Y(t)]_0^\infty \qquad \text{(A-17)}$$

The second term vanishes at both limits, and $\{\cdots\}$ in the first term is the probability density function (pdf) for death at age t, hence I is the mean longevity for the typical ($\mu = 1$) entity. Our argument fails only when $Y(t)$ decays so slowly, e.g. as t^{-2}, that the average age is not finite, which was discussed above.

In summary, one loosely worded sufficient condition for our predictor is the following:

- Specimens have a finite mean lifespan.
- Among individuals, expectations vary by a random time-compression/expansion multiplier (μ).
- That multiplier is uniformly distributed from zero to a maximum that exceeds other rates in the problem at hand.

Given the impressive fits to statistical data in Section 2.2, other sufficient conditions for our predictor probably exist.

Appendix B

Posterior survivability

To obtain a formula for posterior survivability G, we calculate the formula for $Q(T)$ in two different ways and then compare the results. The first way jumps directly from time zero to T, which is Equation 1. The second way inserts an intermediate time at which an observer determines the entity's age A and inquires about it future $F = T - A$, or $T = A + F$.

Now consider a big statistical ensemble of newborn entities. A fraction $Q(T) = Q(A + F)$ of them survives both intervals. Next let us calculate that same fraction one part at a time. The probability of surviving for A is $Q(A)$, and the fraction of that fraction that continues to survive the second interval from A to $A + F$ is $G(F \mid A)$. Hence, the fraction surviving both intervals is the product, $Q(A) \cdot G(F \mid A)$. Equate the two expressions for $Q(T)$, the final fraction:

$$Q(A + F) = Q(A) \cdot G(F \mid A) \tag{B-1}$$

Solve for G, use Equation 1 for Q, and find the desired formula:

$$G(F \mid A) = \frac{Q(A + F)}{Q(A)} = \frac{J + A}{J + A + F} = \frac{P}{P + F} = \frac{1}{1 + F/P} \quad \text{Copy of Eq. 2} \tag{B-2}$$

For brevity the fourth expression changes $A + J$ to P for past. If a case arises where J represents gestation exclusively, then past P is lifetime measured from conception, not birth, which would be A.

#   #   #

Quite often probability theorists write the prior simply as

$$Q = \text{constant}/T \tag{B-3}$$

where "constant" means any number that does not change with time, typically 1.0, and T is the appropriate time. They do not worry about either the infinity at $T = 0$ nor the vague numerator because they know to look ahead a few steps when both will disappear from the posterior probability during the following process. If $T = A + F$, the posterior based on Equation B-3 becomes

$$G = \frac{Q(A+F)}{Q(A)} = \frac{\text{constant}}{A+F} \cdot \frac{A}{\text{constant}} = \frac{A}{A+F} = \frac{1}{1+F/A} \tag{B-4}$$

which is Equation 5 again, Gott's predictor for the naive observer who makes no corrections for gestation or even common sense.

Appendix C

Infinite mean duration

For Gott's survival predictor,

$$G = 1/(1 + F/P) \qquad (2)$$

the probability density of expiring at future time F is:

$$\text{Hazard rate:} \quad r(F) = -dG/dF$$

and so the mean future is

$$\langle F \rangle = \int_0^\infty F \cdot r(F)\, dF = -\int_0^\infty F \frac{dG}{dF}\, dF$$

Integrating by parts,

$$\langle F \rangle = \int_0^\infty G\, dF + GF \Big|_0^\infty$$

The second term is finite:

$$\frac{F}{1 + F/P} \bigg|_0^\infty = P$$

But the first one is not:

$$\int_0^\infty G\, dF = \int_0^\infty \frac{dF}{1 + F/P} = P \int_0^\infty \frac{dx}{1 + x} = P \cdot \ln(1 + x) \Big|_0^\infty = \infty \qquad \text{(C-1)}$$

And so the mean is $\langle F \rangle = \infty$, which compares to the median $F = P$. However, Equation C-1 shows that the integral diverges logarithmically, the slowest possible way. Hence, the slightest long-term decline in the entity's vitality will produce a finite mean. The empirical examples in Section 2.2 and 2.3 suggest that this always happens, and so the infinite mean cannot be considered a reason to reject the theory. See also the discussion of Equations D-28 and D-29 in Appendix D.

One normally thinks of mean and median as similar—sort of middling. But consider a set of 9 objects in which 8 of them weigh 1 gram, and one weighs 1 kilogram. The median (fifth) weight is 1 gram, and the mean is about $1008/9 = 112$ grams. Businesses, stage productions, and species are all like that. A few big winners go on and on inflating the mean but not the median.

Appendix D

Survival predictor from Bayes' theory

A conventional derivation of Gott's survival predictor provides reassuring closure with established techniques. It uses Bayes' theorem, which begins with two different expressions for the joint probability of any two random variables X and Y, symbolized $\text{Prob}(X, Y)$. First it is expressed as the prior probability of Y times the conditional probability of X given Y. Then the second form simply exchanges X and Y:

$$\text{Prob}(X, Y) = \text{Prob}(X \mid Y) \cdot \text{Prob}(Y)$$
$$= \text{Prob}(Y \mid X) \cdot \text{Prob}(X) \qquad \text{(D-1)}$$

A word of caution: this is not proper notation for functions. Normally "Prob" would be shorthand for some formula, and $\text{Prob}(Y)$ would be the output you get by putting Y into that formula. Then $\text{Prob}(X)$ would be the output you get by putting X into that *same* formula, the one called "Prob". Here $\text{Prob}(X)$ and $\text{Prob}(Y)$ are two *different* formulas, the former giving the probability of X, the latter giving the probability of Y. This simplified notation would cause confusion if we wanted to evaluate $\text{Prob}(7.6)$. Which formula applies? However, this ambiguity does not arise in the following context, so let us keep the notation simple.

The joint probability at the left end of Equation D-1 has already served its purpose. From now on we equate only the two terms with conditional probability:

$$\text{Prob}(X \mid Y) \cdot \text{Prob}(Y) = \text{Prob}(Y \mid X) \cdot \text{Prob}(X) \qquad \text{(D-2)}$$

One of these conditional probabilities is known and used to evaluate the other.

In a typical application we observe an effect E, which may be attributed to any one of n possible causes C_j, and now we want to revise the probability of each cause based on our observation of E. Put $X = C_j$, $Y = E$, and solve for the desired probability of C_j, which in this context is called the likelihood of C_j:

$$\text{Prob}(C_j \mid E) = \frac{\text{Prob}(C_j)}{\text{Prob}(E)} \text{Prob}(E \mid C_j) \tag{D-3}$$

The prior probability $\text{Prob}(E)$ can be expressed in terms of the possible causes:

$$\text{Prob}(E) = \text{Prob}(E \mid C_1) \text{Prob}(C_1) + \text{Prob}(E \mid C_2) \text{Prob}(C_2) + \cdots$$
$$+ \text{Prob}(E \mid C_n) \text{Prob}(C_n) \tag{D-4}$$

Substitution in Equation D-3 gives

$$\text{Prob}(C_j \mid E) = \frac{\text{Prob}(E \mid C_j) \text{Prob}(C_j)}{\sum\limits_{i=1}^{n} \text{Prob}(E \mid C_i) \text{Prob}(C_i)} \tag{D-5}$$

which is Bayes' theorem in its usual form. As a sanity check, note that the sum of all the likelihoods must equal 1.0, and indeed that sum makes the numerator and denominator identical on the right side of Equation D-5.

Suppose we randomly observe ages from a long-lived population. On average (but not in every instance) we certainly expect these ages to exceed those drawn from a short-lived population. Thus we can think of duration T as a cause and observed ages A as an effect and apply Bayes to find the probability of T given A and thereby prove Gott's predictor. In our case T and A are continuous variables, whereas C and E above are discrete, but the conversion is straightforward.

Our problem is to prove Gott's predictor, Equation 5, using the probability of age given duration, which is obvious from Figure 7 in Section 2.1:

$$H(A \mid T) = A/T \tag{D-6}$$

Putting $A + F = T$ in Equation 5 gives the same result, which is what we strive to prove using Bayes:

$$G(F \mid A) = \frac{A}{A + F} = \frac{A}{T} \tag{D-7}$$

Unfortunately we cannot plug G and H directly into Bayes' theorem. To see why, let us denote *probability densities* as $Pt(t)$ for duration and $Pa(a)$ for age. (The double t and double a appear to be redundant, but strange as it seems, we shall encounter $Pa(t)$ below.) Next let us write out the full definitions of G and H in terms of inequalities:

$$\left. \begin{array}{l} G = \text{Prob}(\text{duration} > T \mid A) = \displaystyle\int_{T}^{\infty} Pt(t \mid A) \, dt \\[2ex] H = \text{Prob}(\text{age} < A \mid T) = \displaystyle\int_{0}^{A} Pa(a \mid T) \, da = A/T \end{array} \right\} \tag{D-8}$$

Compare Equation D-6.

There is no way to relate these two integrals via Bayesian equations because they apply to different variables over different ranges. However, there is a way to avoid the integrals and apply Bayes to the infinitesimal probabilities in an "area" dT by dA.

Differentiating Equations D-8 gives the required probability densities, first the one for H:

$$\partial H/\partial A = Pa(A\,|\,T) = 1/T \tag{D-9}$$

which represents the uniform distribution of ages over duration T, in other words the derivative of Equation D-6. The corresponding density for G is

$$-\partial G/\partial T = Pt(T\,|\,A) \tag{D-10}$$

To interpret this, think of a big statistical ensemble. The left side is the rate at which the fraction surviving decreases, which obviously is the fractional rate at which entities are expiring.

Now Bayes' theorem, Equation D-1, applies to infinitesimal probabilities within both dT and dA:

$$Pd(T,A)\,dT\,dA = [Pa(A)\,dA][Pd(T\,|\,A)\,dT] = [Pt(T)\,dT][Pd(A\,|\,T)\,dA]$$

The infinitesimals cancel out leaving

$$Pt(T,A) = Pa(A) \cdot Pt(T\,|\,A) = Pt(T) \cdot Pa(A|T) \tag{D-11}$$

Use Equations D-10 and D-11 to solve for Gott's predictor:

$$-\frac{\partial G}{\partial T} = \frac{Pt(T)}{T \cdot Pa(A)} \tag{D-12}$$

To get $Pa(A)$, recall Equation D-5; for which we need the analogous equation for the case of continuous variables, namely,

$$Pa(A) = \int_A^\infty Pa(A\,|\,T) \cdot Pt(T)\,dT = \int_A^\infty Pt(T)\frac{dT}{T} \tag{D-13}$$

Now the solution for G, Equation D-12, depends entirely on $Pt(T)$. So far the theory has been rigorous, but it gets fuzzy when we try to choose a plausible probability density $Pt(T)$ for durations prior to any observation of the entity in question. There are guidelines for this, especially a so-called noninformative (or vague) prior, which serves to express complete ignorance about the entity in question prior to observing it. On the interval zero to infinity the usual noninformative prior [77] is simply

$$Pt(T) = 1/T \tag{D-14}$$

The fact that this is not a proper probability distribution does not matter as discussed below Equation B-3 in Appendix B and again below. Use this in Equation D-13 and find

$$Pa(A) = 1/A \tag{D-15}$$

(As a sanity check, note that we could have written this equation directly as the noninformative prior applied directly to A. This agreement tends to substantiate the idea of the noninformative prior.)

Equation D-12 now gives

$$-\frac{\partial G}{\partial T} = \frac{Pt(T)}{T \cdot Pa(A)} = \frac{A}{T^2}; \qquad G = \frac{A}{T} = \frac{A}{A+F} = \frac{1}{1+F/A} \qquad \text{(D-16)}$$

which is Gott's original predictor, the desired result. Although Bayes' theorem and the uninformative prior are widely used, they have logical weaknesses and are not universally accepted. One might say that GSP substantiates Bayes rather than the converse.

We could stop here, but it is instructive to look at priors for cases where we do know something about the entity in advance.

D.1   ADVANCE KNOWLEDGE

Suppose we have an estimate of Q with a gestation period J and perhaps an additional term for long-term decay rate that represents obsolescence. In this case a good sanity check is to solve the equations for the prior probabilities of age and duration and then compare their features to insure that they are reasonable.

Since differentiation is often easier than integration, let us use the differential form of Equation D-13. With x as a dummy variable, it becomes

$$Pt(x) = -x\frac{d}{dx}Pa(x) \qquad \text{(D-17)}$$

As practice for the next step, let us prove something already known: $Pt(x)$ and $Pa(x)$ have equal normalizations Nt and Na, usually 1.0.

$$Nt = \int_0^\infty Pt(t)\, dt = -\int_0^\infty x\frac{d}{dx}Pa(x)\, dx$$

Integration by parts yields

$$Nt = -[xPa(x)]_0^\infty + \int_0^\infty Pa(x)\, dx = 0 + Na$$

The [square bracket] vanishes at ∞ if $Pa(x)$ decays faster than $1/x$, which it always does; otherwise, normalization does not exist, which then proves that the two normalizations are equal.

Let us now repeat this process to compare mean age $\langle A \rangle$ to mean duration $\langle T \rangle$:

$$\langle T \rangle = \int_0^\infty tPt(t)\, dt = -\int_0^\infty t^2\left(\frac{d}{dt}Pa(t)\right) dt$$

$$= [t^2 Pa(t)]_0^\infty + \int_0^\infty 2tPa(t)\, dt = 2\langle A \rangle$$

That is,

$$\langle A \rangle = \langle T \rangle/2 \qquad \text{(D-18)}$$

If observers arrive at random times in the duration of an entity and determine its age, then their average arrival time is half the duration, and so Equation D-18 reassures us that the theory is working.

Substitute Equation D-17 into D-12 and find

$$\frac{\partial}{\partial T} G(T \mid A) = \frac{\frac{d}{dT} Pa(T)}{Pa(A)}$$

Integrate and find that $G = Pa(T)/Pa(A) + K$. But when future $F = 0$, $T = A$, $G = 1$, hence the constant $K = 0$. Therefore,

$$G(T \mid A) = \frac{Pa(T)}{Pa(A)} \tag{D-19}$$

Recall how G relates to the original prior survivability functions $Q(T)$, essentially Equation B-1:

$$G(T \mid A) = \frac{Q(T)}{Q(A)} \tag{D-20}$$

Compare to Equation D-19. Since A and T are independent, the only solution is

$$Pa(x) = CQ(x) \tag{D-21}$$

where x is a dummy variable that represents either duration T or age A, and the constant C has dimensions of inverse time.

#   #   #

Let us examine one practical example of the prior probabilities Pa and Pt just to verify that they behave as we might expect. We start with $Q(T)$, a survivability curve of the sort shown in Section 2.3. A practical form is

$$Q(T) = \frac{J}{J + T + \eta \cdot T^a} \tag{D-22}$$

Here J is the so-called "gestation" period in Equation 1 (which is actually the reciprocal of the cutoff hazard rate). The η term in the denominator causes G to decay faster after a time $T = 1/\eta^{1/(a-1)}$. On a log–log plot of Q versus T like those in Section 2.3, this term causes the slope to bend from -1 to $-\alpha$, which is roughly what we find for microcosm statistics in Figures 9 through 16 with α in the range 2 to 3. This slope change represents imperfect statistical indifference. If you observe an entity and recognize that it is very old, then you think it is more likely in the last quarter of its life than the first because some process causes aging or obsolescence. Hence, practical GSP decays a bit faster than the ideal in Equation 1.

Now Equation D-21 gives the prior for age:

$$Pa(A) = C \cdot Q(A) \tag{D-23}$$

and C is the constant that normalizes Pa:

$$C = 1 \bigg/ \int_0^\infty Q(x)\, dx \tag{D-24}$$

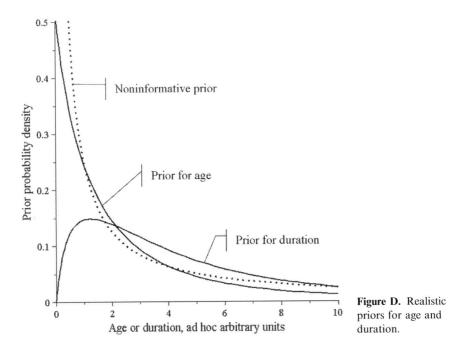

Figure D. Realistic priors for age and duration.

Equation D-17 gives the prior probability for duration:

$$Pt(T) = -C \cdot T \frac{d}{dT} Q(T) \tag{D-25}$$

Finally, Figure D shows plots of Equations D-23 and D-25 with $\eta = 0.1$ and $\alpha = 2.5$, the final slope in Figure 16. Equation D-24 gives $C = 0.516$. For comparison Figure D also shows the ideal curve (dotted) for the case of no obsolescence and an observer who knows absolutely nothing about the process, namely,

$$Pt(x) = Pa(x) = 0.25/x \tag{D-26}$$

The prior density for age peaks at $A = 0$ because an observation immediately after birth is possible regardless of the unknown duration, whereas observation at an old age is possible only if duration is even longer. The prior for duration goes to zero at $T = 0$ because the benefit of gestation and/or a maximum hazard rate prevents the process from expiring at the first instant after birth. (A stage production does not expire as the first curtain rises.)

It looks plausible that $\langle A \rangle = \langle T \rangle / 2$ in Figure D, which was proven in general in the steps above Equation D-18. In fact, that equation is the most convincing output of this whole exercise because common sense demands that it be exactly true: the observer is just as likely to arrive in the first half of the duration as in the second.

D.2   NONINFORMATIVE PRIOR PROBABILITY

The noninformative prior on the interval 0 to ∞ is worthy of additional comment. It applies to the situation in which we know nothing about the process in question. This prior cannot have parameters. For example, if the random variable is x, then the expression

$$\frac{2a}{\pi} \frac{1}{a^2 + x^2}$$

is a perfectly good probability density in some other problem, but not here because it contains parameter a. If a is a property of the process, then knowing it would violate the assumption that we know nothing. Otherwise, a would have to be some universal constant, but there is no hint that any such constant exists. If it did, it would have to apply to all scales, say lifetimes of everything from insects to mountains. Hence, for lack of parameters the noninformative prior must be something very simple, namely a power law:

$$\text{Noninf}(x) = C \cdot x^p \tag{D-27}$$

If $p = -1$, this prior is improper because its integral

$$\int_0^\infty \text{Noninf}(x)\, dx = C \int_0^\infty \frac{dx}{x} = \ln(x) \Big|_0^\infty \tag{D-28}$$

is infinite at both limits, 0 and ∞. However, it is a very soft infinity. To make it finite, simply replace x by the average of $x^{1+\varepsilon}$ and $x^{1-\varepsilon}$, where ε is very small. Then putting $C = \varepsilon/\pi$ gives a perfectly proper normalized probability density no matter how small ε may be:

$$\text{Noninf}(x) = \frac{\varepsilon}{\pi} \frac{2}{x^{1+\varepsilon} + x^{1-\varepsilon}}; \qquad \int_0^\infty \text{Noninf}(x)\, dx = 1 \tag{D-29}$$

Here ε looks suspiciously like a parameter, but not really because we can simply choose ε so small that it has no practical effect in the problem at hand. Then the problem of an improper prior is solved. Thus we may as well regard C/x as a legitimate probability density in problems where C cancels out in a ratio as it does when Equation B-3 is used in B-4.

   Not so for other powers p in Equation D-10. If we test $p > -1$ in Equations D-27 and 28 it gives a hard infinity at the upper limit, or if $p < -1$, it gives a hard infinity at the lower limit, neither of which is acceptable. Therefore, Jeffreys' choice of $1/x$ for prior probability density is surely the leading candidate for noninformative prior on the interval 0 to ∞. Moreover, since this prior integrates to the logarithm, Equation D-28, this says that all orders of magnitude are equally probable (as are octaves, nepers, or any other interval based on powers of some number). This seems a very convincing and satisfying way of saying that we know nothing about the process in question.

Appendix E

Stage productions running on specified dates

Here we develop the theory that produced Figure 16, the survivability Q of stage productions derived from the statistics of shows playing on specified dates. In preparation let us examine the rate at which new shows open. This rate may change during the period in question. First, we must decide what shows to count. Many of them are one-night stands and performances by groups that specialize in filling gaps in the theaters' calendar. Although they surely hope for a big hit, some are not serious competitors for the long runs. Hence, we should drop the latter from the statistics, but we have no conclusive criterion to distinguish them from the mainstream. Let us test two cases, one that counts all shows, and one that counts all except the single performances.

Figure E shows the cumulative number of starts counting from January 1890 through December 1959, the full range of Wearing's calendar of stage shows. The slopes of these curves are the start rates, which are constant throughout the period of main interest, the twenties and early thirties. A straight line would fit rather well over the full range including wartime, and its slope would be only slightly less. World War II hit the theater much harder than WWI even though the Spanish flu also hit in 1918. The hump during the 1890s is probably an artifact because some of these productions actually opened during the 1880s, but I have no data prior to 1890 with which to make the correction.

In the end we obtain an effective start rate defined as the value that normalizes Q, meaning $Q(0) = 1.0$ and $Q(\infty) = 0$. This rate is only 131 shows/year, which says that we should omit some two-night stands as well. Whatever the exact value of the slope, its constancy is very fortunate for data reduction. The theory below requires a start rate at the time each show opened. Had it not been constant, I would have been forced to trace each show back to its beginning and look up the rate at that time.

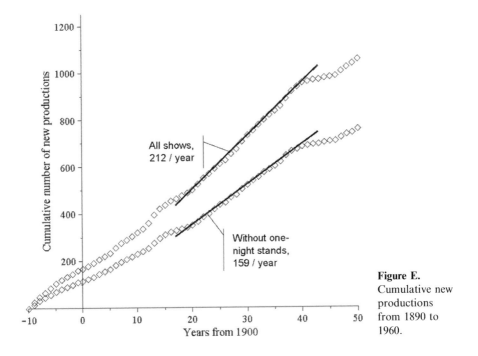

Figure E. Cumulative new productions from 1890 to 1960.

Let p denote the statistic that we take from Wearing's data for each show open on the given date. It could denote any of three quantities:

- the number of performances prior to the given date
- the number after that date, or
- the total number, prior plus future.

I chose the third because Wearing or somebody has already done the hard work of tallying the total from the detailed list of performance dates.

Let $S(p)$ denote the number of shows playing on the given date that will ultimately survive for at least p more performances. Let t denote time *prior* to the given date; in other words greater t is earlier time. Let r denote the number of performances per day on average. Let B (for birthrate) denote the constant start rate discussed above. Then for productions starting during the interval dt, the number surviving on the given date is

$$dS_1 = B \cdot Q(rt)\,dt, \quad \text{when } rt > p$$

When $rt < p$, the show must survive longer to qualify for the count S:

$$dS_2 = B \cdot Q(p)\,dt, \quad \text{when } rt < p$$

Integrating t over both ranges gives

$$S(p) = \frac{B}{r}\left(pQ(p) + \int_p^\infty Q(y)\, dy\right) \tag{E-1}$$

The goal is to relate increments of S, namely single performances, to increments of Q and then compare the result to graphs like Figure 15 and thereby extend our data to long-running shows. Toward that end, differentiate Equation E-1 with respect to p and simplify. The result is simply

$$dS = (Bp/r)\, dQ \tag{E-2}$$

(This would have been something complicated had B not been constant.) We take increments in S to be single performances listed by Wearing, and sum the corresponding increments of Q on a spreadsheet to get $Q(p)$. It works best to begin with the longest running show and work backward to $p = 0$. For a single date we would therefore use $dS \rightarrow \Delta S = -1$, but since we are summing five dates for smoother data, we must undo that multiplier and use

$$\Delta S = -1/5 \tag{E-3}$$

Solving Equations E-2 and E-3 for ΔQ gives

$$\Delta Q = -\frac{r}{5B \cdot p} \tag{E-4}$$

which is summed to Q using a spreadsheet.

   The results are plotted in Figure 16 in Section 2.3. Theaters and performances on the five dates appear in Table E-1 below. The spreadsheet in Table E-2 lists Wearing's raw data and also tallies Q from them. Certain long-running shows do not fit on this spreadsheet. Table E-2 contains only the summary tally for them, while the details appear in Table E-3.

Table E-1. Stage productions playing on five specific dates in London.

Theater	Fri. 17 Aug 23	Wed. 04 Jun 24	Mon. 23 Mar 25	Sat. 09 Jan 26	Thur. 28 Oct 26
Adelphi	Young Person in Pink	Diplomacy	Iris	Betty in Mayfair	Merely Molly
Aldwych	Tons of Money	It pays to Advertise	It Pays to Advertise	Cuckoo in the Nest	Rookery Nook
Ambassadors	Lilies of the Field	Collusion	Any House	Madrass House	Escape
Apollo	What Every Woman Knows	The Fake	By the Way	Tricks	Fall Guy
Comedy	Peace and Quiet		The Vortex	9.45 [sic]	After Dark
Court	Omar Khayyam	Farmer's Wife	Farmer's Wife	Farmer's Wife	Farmer's Wife
Criterion		The Mask and the Face	Just Married	Hay Fever	Scarlet Lady
Daly's	Merry Widow	Madame Pompadour	Dollar Princess	Katja	Yvonne
Drury Lane		London Life	Rose Marie	Rose Marie	Queen was in the Parlor
Duke of York's		Punch Bowl	Punch Bowl	No. 17 [sic]	Lady, Be Good !
Empire	Enemies of Women		Boodle	Henry VIII	Rat Trap
Everyman	Mary Stuart		Painted Swan	Inheritors	
Gaiety		Our Nell		Blue Kitten	
Garrick		Rising Generation	Katja	Blue Bird	Ghost Train
Globe	Bluebeard's Eighth Wife	Our Betters	Possessions	Lullaby	Ask Beccles
Haymarket	Prisoner of Zenda	Great Adventure	Grand Duchess	Man with Load of Mischief	None but the Brave
Hippodrome	Brighter London	Leap Year	Hamlet	Mercenary Mary	Yellow Sands
King's		Merry Wives of Windsor	Better Days (twice daily)		
Kingsway		Yoicks !	Are You a Mason ?	The Old Adam	Rosmersholm
Little Theatre	Nine O'clock Revue	2nd Little Revue Starts at 9	Persevering Pat	Too Much Money	London's Potiniere Revue
London Pavilion	Dover Street to Dixie	Ten Commandments	North of 36	Still Dancing	Black-Birds
Lyceum		Merry Widow		Panto: Dick Whittington	The Padre
Lyric Theatre	Lilac Time	Way of the World	Street Singer	Lilac Time	Best People
New	Eye of Siva	Saint Joan	The Tyrant	Quinney's	The Constant Nymph
New Oxford's Theatre	Little Nellie Kelly	L' Aventuriere	Khaki (twice nightly)	Alf's Button	
Palace Theatre	Music Box Revue		No No Nanette	No No Nanette	Princess Charming
Palladium		Whirl of the World	Sky High	Cinderella	Life

Theater	Fri. 17 Aug 23	Wed. 04 Jun 24	Mon. 23 Mar 25	Sat. 09 Jan 26	Thur. 28 Oct 26
Playhouse	Enter Kiki !	White Cargo	White Cargo	A Doll's House	Romance
Prince of Wales	So This is London !	Seven Who Were Hanged	Charlot's Revue	Charley's Aunt	Charlot Show of 1926
Queen's	Stop Flirting	Elsie Janis at Home	Dancing Mothers	Man in Dress Clothes	And So to Bed
Regent	Robert E. Lee	Romeo and Juliet	Saint Joan	Androcles and the Lion	
Royalty	At Mrs. Beams	Bachelor Husbands	The Pelican	Juno and the Paycock	The Lash
Savoy	Polly (seq. Beggar's Opera)	The Lure	Sport of Kings	The Unfair Sex	Love's a Terrible Thing
Shaftesbury		Toni	Lightnin'	Peter Pan	Just a Kiss
St. James's	The Outsider	Green Goddess	Grounds for Divorce	Last of Mrs. Cheyney	Last of Mrs. Cheyney
St. Martin's	The Will & Likes of Her	In the Next Room	Spring Cleaning	Ghost Train	Berkeley Square
Strand		Stop Flirting	Patricia	Treasure Island	Whole Town's Talking
Vaudeville	Rats !	"Puppets"	Fata Morgana		R.S.V.P.
Winter Garden		To-Night's the Night	Primrose	Tell Me More	Tip-Toes
Wyndham's	The Dancers	To Have the Honour	Man with a Heart	Rising Generation	The Ringer
His Majesty's			Bamboula	Co-Optimists	Co-Optimists
Old Vic			Macbeth	Bohemian Girl	Henry V & Aida
Fortune				Are You a Mason?	A Month in the Country
Prince's Theatre				White Cargo	The Gondoliers
Prince's (matinee)				When Knights were Bold	
Scala				Don't Tell Timothy	
Lyric, Hammersmith			The Rivals		Riverside Nights

The following Theaters are not included in Wearing's books.

Theater	Fri. 17 Aug 23	Wed. 04 Jun 24	Mon. 23 Mar 25	Sat. 09 Jan 26	Thur. 28 Oct 26
Barnes				Ivanov	Three Sisters
Century				Jean Stirling Mackinlay	
Chelsea Palace				Bluebell in Fairyland	Doctor's Dilemma
Holborn Empire				Where the Rainbow Ends	
The Q				Tame Cat	Habit
Victoria Palace				Windmill Man	

Table E-2. Spreadsheet for computing survivability $Q(p)$.

Short Name	L.T. Date	Run Open	Run Close	Prfs/ Day	W'ch Run	1st Prfs	2nd Prfs	3rd Prfs	4th Prfs	5B·ΔQ	5B·Q	Total Perfs	
Footnotes	a			b	c			d		e	e		
7 who were Hanged	2	28-Apr-24	[4 Jun 24]	0.94	1	12	2			0.0668	1.4154	14	
Rat Trap, The	5	18-Oct-26	5-Nov-26	1.00	1	19				0.0526	1.3486	19	F
Anyhouse	3	12-Mar-25	28-Mar-25	1.18	1	20				0.0588	1.2960	20	o
Painted Swan	3	16-Mar-25	4-Apr-25	1.05	1	21				0.0500	1.2371	21	o
Inheritors	4	28-Dec-25	16-Jan-26	1.05	1	21				0.0500	1.1871	21	t
Peace & Quiet	1	31-Jul-23	18/Aug/23	1.16	1	22				0.0526	1.1371	22	n
Man with a Heart	3	14-Mar-25	4-Apr-25	1.14	1	25				0.0455	1.0845	25	o
LondnPotiniereRevue	5	8-Oct-26	30-Oct-26	1.13	1	26				0.0435	1.0391	26	t
Love'sTerribleThing	5	4-Oct-26	30-Oct-26	1.15	1	31				0.0370	0.9956	31	e
Possessions	3	23/Mar/25	18/Apr/25	1.15	2	3	31			0.0338	0.9585	34	s
London Life	2	3-Jun-24	5-Jul-24	1.18	1	39				0.0303	0.9248	39	
Bachelor Husbands	2	2-Jun-24	28/Jun/24	1.19	1	32	8			0.0296	0.8945	40	
Fall Guy, The	5	20-Sep-26	30-Oct-26	1.17	1	48				0.0244	0.8648	48	
Lash, The	5	26-Oct-26	4/Dec/26	1.18	1	47	12			0.0199	0.8405	59	
Tricks !	4	22-Dec-25	13-Feb-26	1.13	1	61				0.0185	0.8205	61	
Lure, The	2	8/May/24	21/Jun/24	1.16	2	1	52	12		0.0178	0.8020	65	
Old Adam, The	4	17-Nov-25	16-Jan-26	1.10	1	67				0.0164	0.7842	67	
After Dark	5	20-Sep-26	6/Nov/26	1.17	1	56	12			0.0172	0.7679	68	6
Dollar Princess	3	4-Feb-25	4-Apr-25	1.15	1	69				0.0167	0.7507	69	
Don't Tell Timothy	4	15-Dec-25	27-Feb-26	1.00	1	71				0.0141	0.7340	71	
ElsieJanis at Home	2	2-Jun-24	9-Aug-24	1.04	1	72				0.0145	0.7199	72	
Bamboula	3	24-Mar-25	30-May-25	1.13	1	77				0.0147	0.7054	77	
Scarlet Lady, The	5	30-Sep-26	11-Dec-26	1.15	1	84				0.0137	0.6907	84	
Merely Molly	5	22-Sep-26	4-Dec-26	1.15	1	85				0.0135	0.6770	85	F
Omar Khayyam	1	21-Aug-23	3/Nov/23	1.15	1	86				0.0133	0.6635	86	o
Charlot Show, 1926	5	5-Oct-26	18-Dec-26	1.16	1	87				0.0133	0.6502	87	o
Lullaby	4	6-Nov-25	23-Jan-26	1.11	1	88				0.0127	0.6369	88	t
Eye of Siva	1	8-Aug-23	20/Oct/23	1.16	1	86	6			0.0126	0.6242	92	n
Just a Kiss	5	8-Sep-26	27-Nov-26	1.15	1	93				0.0123	0.6116	93	o
Boodle	3	10-Mar-25	30-May-25	1.15	1	94				0.0122	0.5992	94	t
Too Much Money	4	26/Dec/25	30/Jan/26	1.17	2	62	42			0.0112	0.5870	104	e
None but the Brave	5	30-Jul-26	30-Oct-26	1.14	1	106				0.0108	0.5758	106	s
Madras House	4	30/Nov/25	27/Feb/26	1.08	2	10	97			0.0101	0.5651	107	
DoverStreet to Dixie	1	31-May-23	1/Sep/23	1.15	1	108				0.0106	0.5550	108	
Robert E. Lee	1	20-Jun-23	22/Sep/23	1.15	1	109				0.0105	0.5444	109	
Mary Stuart	1	30/Jul/23	18/Aug/23	1.10	2	69	22	19		0.0100	0.5338	110	
Grand Duchess	3	20-Feb-25	4/Apr/25	1.14	1	50	60			0.0103	0.5238	110	
Dancing Mothers	3	17-Mar-25	20-Jun-25	1.15	1	110				0.0104	0.5135	110	
Still Dancing	4	19-Nov-25	27-Feb-26	1.13	1	114				0.0099	0.5031	114	
Music Box Revue	1	15-May-23	18/Aug/23	1.25	1	120				0.0104	0.4932	120	
Tyrant, The	3	18-Mar-25	4-Jul-25	1.16	1	126				0.0092	0.4828	126	
Grounds for Divorce	3	21-Jan-25	2/May/25	1.14	1	116	12			0.0089	0.4736	128	
Better Days	3	19-Mar-25	6-Jun-25	1.69	1	135				0.0125	0.4647	135	8
Nine.45 [9.45]	4	22-Dec-25	10/Apr/26	1.13	1	124	6	6		0.0083	0.4522	136	
Queen in the Parlour	5	24-Aug-26	18-Dec-26	1.17	1	137				0.0085	0.4439	137	F
Blue Kitten, The	4	23-Dec-25	24-Apr-26	1.15	1	141				0.0081	0.4354	141	o
Our Nell	2	16-Apr-24	16/Aug/24	1.14	1	140	8			0.0077	0.4272	148	o
Collusion [Terry]	2	1-Apr-24	9-Aug-24	1.15	1	150				0.0076	0.4195	150	t
Iris	3	21/Mar/25	1/Jul/25	1.46	2	3	150			0.0095	0.4119	153	n
Enter Kiki !	1	2-Aug-23	15/Dec/23	1.14	1	155				0.0074	0.4024	155	o
Life [Cowan et al]	5	30-Aug-26	4-Dec-26	1.72	1	167				0.0103	0.3950	167	t
Patricia	3	31-Oct-24	28-Mar-25	1.14	1	170				0.0067	0.3847	170	e
Lightnin'	3	27-Jan-25	6/Jun/25	1.14	1	149	23			0.0066	0.3780	172	s
2nd Li'l Revue..at 9	2	18-Mar-24	16/Aug/24	1.14	1	173				0.0066	0.3714	173	7
Tip-Toes	5	31-Aug-26	12-Feb-27	1.09	1	181				0.0060	0.3648	181	
Betty in Mayfair	4	11-Nov-25	24-Apr-26	1.10	1	182				0.0061	0.3588	182	
Outsider, The	1	31-May-23	1/Sep/23	1.14	1	107	7	6	66	0.0061	0.3527	186	
Padre, The	5	22-May-26	6-Nov-26	1.15	1	194				0.0059	0.3466	194	

Short Name / Footnotes	L.T. Date (a)	Run Open	Run Close	Prfs/Day (b)	W'ch Run (c)	1st Prfs (d)	2nd Prfs	3rd Prfs	4th Prfs	5B·ΔQ (e)	5B·Q (e)	Total Perfs	
To Have the Honour	2	22-Apr-24	11-Oct-24	1.13	1	196				0.0058	0.3407	196	
Ask Beccles	5	20-Jul-26	20/Nov/26	1.15	1	143	42	12		0.0059	0.3349	197	
Whole Town's Talking	5	7-Sep-26	18/Dec/26	1.15	1	118	12	72		0.0057	0.3291	202	1
In the Next Room	2	6-Jun-24	29/Nov/24	1.14	1	202	12			0.0053	0.3234	214	
Charlot's Revue	3	23-Sep-24	28-Mar-25	1.16	1	216				0.0053	0.3181	216	
Are You a Mason?	4	28/Dec/25	30/Jan/26	1.76	3	63	87	60	12	0.0079	0.3127	222	8
Androcles & Lion	4	26/Dec/25	30/Jan/26	1.17			42			0.0052	0.3048	224	13
Berkeley Square	5	6-Oct-26	5/Mar/27	1.19	1	179	47			0.0052	0.2996	226	
Alf's Button	4	17/Dec/25	23/Jan/26	1.66	2	111	63	57		0.0072	0.2943	231	8
Riverside Nights	5	10-Apr-26	6/Nov/26	1.13	1	238	1			0.0047	0.2871	239	
Likes of Her	1	15-Aug-23	1/Mar/24	1.15	1	229	10			0.0048	0.2824	239	
Unfair Sex, The	4	9-Sep-25	20/Mar/26	1.17	1	226	3	12		0.0049	0.2776	241	
Toni	2	12-May-24	13-Dec-24	1.16	1	250				0.0046	0.2728	250	
Puppets !	2	2-Jan-24	25/Jul/24	1.24	1	255				0.0049	0.2681	255	
Primrose	3	11-Sep-24	25-Apr-25	1.12	1	255				0.0044	0.2633	255	
Little Nellie Kelly	1	2-Jul-23	16/Feb/24	1.14	1	263				0.0043	0.2589	263	
No. 17	4	12-Aug-25	13/Feb/26	1.12	1	209	15	12	28	0.0043	0.2545	264	4
Rats !	1	21-Feb-23	22/Sep/23	1.23	1	264				0.0047	0.2503	264	
Tell Me More	4	26-May-25	16-Jan-26	1.12	1	264				0.0042	0.2456	264	
Yoicks !	2	11-Jun-24	14-Feb-25	1.07	1	266				0.0040	0.2414	266	
Fake, The	2	13-Mar-24	13/Sep/24	1.14	1	210	46	12		0.0042	0.2374	268	
Spring Cleaning	3	29-Jan-25	29/Aug/25	1.23	1	262	12			0.0045	0.2331	274	
Escape	5	12/Aug/26	12/Mar/27	1.14			242			0.0041	0.2286	275	13
Yvonne	5	22-May-26	29-Jan-27	1.11	1	281				0.0040	0.2245	281	
Fata Morgana	3	15-Sep-24	28/Mar/25	1.25	1	243	24	21		0.0043	0.2205	288	
Rosmersholm	5	30/Sep/26	13/Nov/26	1.16			52			0.0040	0.2162	291	13
Man..Load..Mischief	4	11/Jun/25	16/Jan/26	1.19	2	2	261	30		0.0040	0.2122	293	
R. S. V. P.	5	23-Feb-26	6-Nov-26	1.14	1	294				0.0039	0.2082	294	
So This is London!	1	11-Apr-23	8/Dec/23	1.15	1	278	7	12		0.0039	0.2043	297	
Mask & the Face	2	27/May/24	13/Dec/24	1.15	2	41	232	28		0.0038	0.2004	301	2
Rising Generation	4	21/Dec/25	30/Jan/26	1.32	2	236	54	12		0.0044	0.1966	302	
Rising Generation	2	3-Dec-23	21/Jun/24	1.17	1	236	54	12		0.0039	0.1922	302	8
Young Person in Pink	1	13/Jul/23	6/Oct/23	1.14	3	1	208	98		0.0037	0.1884	307	
Lilies of the Field	1	5-Jun-23	26/Jan/24	1.14	1	269	38			0.0037	0.1847	307	
Best People [G&H]	5	16-Mar-26	11-Dec-26	1.14	1	308				0.0037	0.1809	308	
Sky High	3	30-Mar-25	26/Sep/25	1.71	1	309				0.0055	0.1773	309	9
Black-Birds	5	11-Sep-26	14/May/27	1.13	1	279	32			0.0036	0.1717	311	3
At Mrs. Beam's	1	2/Apr/23	1/Dec/23	1.15	3	2	21	280	12	0.0036	0.1681	315	
Blue Bird, The	4	19/Dec/25	23/Jan/26	1.67			60			0.0053	0.1644	315	8
Polly	1	30-Dec-22	6/Oct/23	1.16	1	325				0.0036	0.1592	325	
Lady, Be Good !	5	14-Apr-26	22-Jan-27	1.14	1	325				0.0035	0.1556	325	
Doll's House, A	4	20/Nov/25	23/Jan/26	1.00			65			0.0031	0.1521	325	13
Pelican, The	3	20-Oct-24	9/May/25	1.20	1	243	12	72		0.0037	0.1490	327	
Vortex, The	3	25-Nov-24	27/Jun/25	1.13	1	244	12	83		0.0033	0.1453	339	F
Will, The	1	15/Aug/23	1/Mar/24	1.15	2	90	229	24		0.0033	0.1420	343	o
Juno & the Paycock	4	16-Nov-24	8/May/26	1.14	1	198	63	20	64	0.0033	0.1386	345	o
Dancers, The	1	15-Feb-23	15/Dec/23	1.15	1	349				0.0033	0.1353	349	t
Street Singer, The	3	27-Jun-24	5-May-25	1.12	1	349				0.0032	0.1320	349	n
Princess Charming	5	21-Oct-26	3-Aug-27	1.26	1	361				0.0035	0.1288	361	o
Quinney's	4	2/Dec/25	30/Jan/26	1.22	2	286	73	10		0.0033	0.1254	369	t
Hay Fever	4	8-Jun-25	31-Mar-26	1.13	1	337	26	18		0.0030	0.1221	381	e
By the Way	3	22-Jan-25	28/Nov/25	1.12	1	347	37			0.0029	0.1191	384	s
Nine O'clock Revue	1	25-Oct-22	1/Sep/23	1.23	1	385				0.0032	0.1162	385	10
Cuckoo in the Nest	4	22-Jul-25	26/Jun/26	1.11	1	376	12			0.0029	0.1130	388	
Sport of Kings, The	3	8-Sep-24	13/Jun/25	1.14			319			0.0029	0.1101	393	13
Bohemian Girl	4	[7 Jan 26]	9 Jan 26]	0.86			2			0.0021	0.1072	400	12,13
Just Married	3	15-Dec-24	19-Dec-25	1.14	1	423				0.0027	0.1051	423	
Green Goddess	2	6-Sep-23	6/Sep/24	1.13	1	416	12			0.0026	0.1024	428	
Stop Flirting	2	29/Mar/24	13/Dec/24	0.75	2	229	194	7		0.0017	0.0997	430	
Stop Flirting	1	30-May-23	15/Dec/23	1.15	1	229	194	7		0.0027	0.0980	430	15
Way of the World	2	7-Feb-24	28/Jun/24	1.10			158			0.0025	0.0953	436	13

Table E-2 (*cont.*).

Short Name	L.T. Date	Run Open	Run Close	Prfs/ Day	W'ch Run	1st Prfs	2nd Prfs	3rd Prfs	4th Prfs	5B·ΔQ	5B·Q	Total Perfs	
Footnotes	a			b	c	d				e	e		
Mercenary Mary	4	7-Oct-25	18-Sep-26	1.29	1	446				0.0029	0.0928	446	
Rookery Nook	5	30-Jun-26	25/Jun/27	1.13	1	409	12	28		0.0025	0.0899	449	
Month in Country	5	6/Oct/26	30/Oct/26	1.20		30				0.0027	0.0874	452	13
Constant Nymph	5	14-Sep-26	13/Aug/27	1.16	1	387	58	12		0.0025	0.0847	457	
Mme. Pompadour	2	20-Dec-23	31/Jan/25	1.13	1	461				0.0024	0.0822	461	
Man...Dress Clothes	4	3/Nov/25	12/Dec/25	2.05	2	232	82	88	60	0.0044	0.0798	462	
Ringer, The	5	1-May-26	23/Apr/27	1.14	1	408	1	53		0.0025	0.0753	462	5
Ev'ryWomanKnows	1	24/May/23	26/Jan/24	1.15	2	63	284	1	120	0.0024	0.0728	468	
Leap Year	2	20-Mar-24	20/Dec/24	1.71	1	471				0.0036	0.0704	471	
Bluebeard's8thWife	1	26-Aug-22	20/Oct/23	1.14	1	482				0.0024	0.0668	482	
Katja	3	20-Feb-25	12-May-26	1.15	1	514				0.0022	0.0644	514	
Katja	4	21-Feb-25	12-May-26	1.15	1	514				0.0022	0.0622	514	15
Prisoner of Zenda	1	23/Aug/23	17/Nov/23	1.26			110			0.0023	0.0599	541	13
Brighter London	1	28-Mar-23	15/Mar/24	1.70	1	603				0.0028	0.0576	603	
Saint Joan (B.Shaw)	2	26-Mar-24	25/Oct/24	1.14			244			0.0019	0.0548	607	13
Saint Joan (B.Shaw)	3	14/Jan/25	9/May/25	1.14			132			0.0019	0.0529	607	15
To-Night's the Night	2	21/Apr/24	30/Aug/24	1.14	2	460	150			0.0019	0.0510	610	
It Pays to Advertise	2	1-Feb-24	11/Jul/25	1.13	1	598	20			0.0018	0.0491	618	11
It Pays to Advertise	3	1-Feb-24	11/Jul/25	1.13	1	598	20			0.0018	0.0473	618	11
Our Betters	2	12-Sep-23	3/Jan/25	1.14	1	548	12	60		0.0018	0.0455	620	
Punch Bowl, The	2	21-May-24	22/Aug/25	1.21	1	554	76			0.0019	0.0436	630	
Punch Bowl, The	3	21-May-24	22/Aug/25	1.21	1	554	76			0.0019	0.0417	630	15
Whirl of the World	2	14-Mar-24	21/Mar/25	1.70	1	635				0.0027	0.0398	635	
Treasure Island	4	26/Dec/25	23/Jan/26	0.86			25			0.0013	0.0371	659	13
And so to Bed	5	6-Sep-26	18/Jun/27	1.17	1	334	72	323		0.0016	0.0358	729	
No No Nannette	3	11-Mar-25	16/Oct/26	1.14	1	665	115			0.0015	0.0342	780	
No No Nannette	4	11-Mar-25	16/Oct/26	1.14	1	665	115			0.0015	0.0328	780	15
Ghost Train, The	4	23-Nov-25	16/Apr/27	1.28			655			0.0016	0.0313	785	13
Ghost Train, The	5	23-Nov-25	16/Apr/27	1.28			655			0.0016	0.0297	785	15
Tons of Money	1	13-Apr-22	29/Jan/24	1.13			743			0.0014	0.0280	797	13
Rivals, The	3	5/Mar/25	23/May/25	1.16			93			0.0014	0.0266	806	13
Last of Mrs. Cheyney	4	22-Sep-25	18/Dec/26	1.13	1	514	12	17	269	0.0014	0.0252	812	
Last of Mrs. Cheyney	5	22-Sep-25	18/Dec/26	1.13	1	514	12	17	269	0.0014	0.0238	812	15
Yellow Sands	5	3-Nov-26	25/Feb/28	1.28			612			0.0015	0.0224	854	13
Doctor's Dilemma	5	17/Nov/26	18/Dec/26	1.16			37			0.0013	0.0209	865	13
Great Adventure	2	5/Jun/24	18/Oct/24	1.18	2	674	160	20	33	0.0013	0.0195	887	
Diplomacy	2	8/Mar/24	24/Jan/25	1.10	2	462	355	97		0.0012	0.0182	914	
White Cargo	2	15-May-24	15/May/26	1.12			821			0.0011	0.0170	1014	13
White Cargo	3	15-May-24	15/May/26	1.12			821			0.0011	0.0159	1014	15
White Cargo	4	15-May-24	15/May/26	1.12			821			0.0011	0.0148	1014	
RoseMarie [H & H]	3	20-Mar-25	26/Mar/27	1.15	1	851	100	149		0.0010	0.0137	1100	
RoseMarie [H & H]	4	20-Mar-25	26/Mar/27	1.15	1	851	100	149		0.0010	0.0126	1100	15
RoseMarie [H & H]	5	20-Mar-25	26/Mar/27	1.15	1	851	100	149		0.0010	0.0116	1100	
Romance [Sheldon]	5	27/Oct/26	19/Feb/27	1.13	2	1047	131			0.0010	0.0105	1178	
Lilac Time	1	22-Dec-22	21/Jun/24	1.15			628			0.0010	0.0096	1179	13
Lilac Time	4	26/Dec/25	13/Mar/26	1.15			90			0.0010	0.0086	1179	
Farmer's Wife	2	11-Mar-24	29/Jan/27	1.25	1	1324	31	39	13	0.0009	0.0076	1407	
Farmer's Wife	3	11-Mar-24	29/Jan/27	1.25	1	1324	31	39	13	0.0009	0.0067	1407	15
Farmer's Wife	4	11-Mar-24	29/Jan/27	1.25	1	1324	31	39	13	0.0009	0.0058	1407	
Farmer's Wife	4	11-Mar-24	29/Jan/27	1.25	1	1324	31	39	13	0.0009	0.0049	1407	
Dick Whittington	4	26/Dec/25	6/Mar/26	1.72	2/day		122			0.0010	0.0041	1643	13
Co-Optimists	4	26/Aug/25	30/Jan/26	1.27			200			0.0007	0.0030	1812	13
Co-Optimists	5	25/Aug/26	12/Feb/27	1.26			216			0.0007	0.0023	1812	15
When Knights...Bold	4	21/Dec/25	16/Jan/26	0.85			23			0.0004	0.0016	2130	13
Peter Pan	4	17/Dec/25	23/Jan/26	1.32			50			0.0004	0.0012	3250	13
Charley's Aunt	4	22/Dec/25	23/Jan/26	1.70			56			0.0005	0.0008	3768	14

The following nine stage productions lasted much longer than Wearing's listings. They are assigned ranks 1 through 9; hence the ranks in the main table above begin with 10.	L.T. Date			
Cinderella	4			
Merry Widow	1,2			
Romeo & Juliet	2			
MerryWives...Windsor	2			
Hamlet	3			
Macbeth	3			
Henry VIII	4			
Henry V	5			
Aida	5			

The following eight shows are missing from Wearing's listing mostly because the theaters did not meet his criteria that year.				
Enemies of Women	1			
French Season	2			
Habit	5			
Presevering Pat	3			
North of 36	3			
Ten Commandments	2			
Khaki	3			
Tame Cat	4			

Footnotes				
a	London Times			
b	Performances per day			
c	Which of the runs occurred on the L.T. date			
d	Italic numbers are perfs for listed date only. For details see Table E-3			
e	See equations, especially E-4.			
1	The 1937 production of Whole Town's Talking was modified to a musical comedy, Oh! You Letty.			
2	Not the same show as "Masks and Faces."			
3	No relation to "Blackbirds" 34.372, 34.255, 36.230, or 36.336.			
4	No clue to run in 1926. Used 15 as average of revivals.			
5	One-night stand assumed, typical of London Repertory Company.			
6	Est. 12 perfs. for 10-day run.			
7	Independent of "The Nine O'clock Revue" except same author.			
8	Two performances per day.			
9	No relation to "Sky High" 42.062 or 48.256.			
10	Independent of "Second Little Revue Starts at Nine" except same author.			
11	Twenty performances is an estimate.			
12	Estimate of performances is based on fragmentary data.			
13	Details appear in the table of Extra-Long Runs, next.			
14	Initial value for 5B-Q is a guestimate.			
15	There are 19 duplicates out of 165 distinct shows. This is 12%.			

Table E-3. Longest-running productions open on selected dates.
Runs listed by Wearing's serial number with number of performances.

	Peter Pan	Charley's Aunt	Co-Optimists	Doll'sHouse	Knights..bold	The Rivals
Title						
SampleDate/Genre	4 / Fairy play, 3 to 5 acts	4 / farcical comedy, 3 acts	4, 5	4 / Play, 3acts	4 / Farce, 3acts	3 / C, usu.3a
First performed	London, 1904, the first run below	Translation 1889	--	Xlation: 6/1889	Nott'ham,9/1906	Jan 1775
Notes, whatever		Orig.Copenhagen,1879	--			
Author	J. M. Barrie, later with John Crook	Brandon Thomas		Henrik Ibsen	C. Marlowe	R.B.Sheridan

Peter Pan Serial #	Prfs	Serial #	Prfs (Continued)	Serial #	Prfs	Charley's Aunt Serial #	Prfs	Serial #	Prfs	Co-Optimists Serial #	Prfs	Doll'sHouse Serial #	Prfs	Knights Serial #	Prfs	The Rivals Serial #	Prfs
04.265	150	30.442	29	47.330	68	92.359	1469	22.337	80	21.180	498	91.017	1	07.026	579	95.248	42
05.310	107	31.531	29	48.329	69	00.227	24	23.285	56	22.257	232	91.168	1	10.003	146	00.036	7
06.310	101	32.460	25	49.321	64	01.152	114	24.404	68	23.242	210	92.102	30	14.160	243	00.048	83
07.364	102	33.388	24	50.315	64	04.242	66	**25.390**	**56**	24.290	209	93.050	14	15.024	24	10.077	1
08.348	82	34.378	29	51.312	61	05.316	46	26.378	56	**25.232**	**200**	01.053	1	17.249	237	15.273	9
09.369	84	35.451	21	52.300	55	07.377	42	28.451	46	26.237	216	03.120	2	19.005	36	16.178	4
10.339	69	36.394	25	53.342	53	08.351	50	29.438	58	29.245	135	08.120	3	20.023	56	20.063	13
11.369	77	37.435	25	54.319	51	09.376	48	30.441	46	30.111	99	11.025	2	20.379	41	24.098	9
12.391	64	38.380	24	55.256	51	10.343	24	33.387	48	35.154	13	11.052	52	21.357	108	**25.054**	**93**
13.403	40	41.154	40	56.288	49	11.374	58	34.376	50			11.327	8	22.352	31	29.075	14
14.345	67	42.175	39	57.296	61	12.393	54	38.374	50			12.079	1	23.282	105	34.008	37
15.327	50	43.183	38	58.334	55	13.399	59	47.332	101			12.324	6	24.401	79	35.377	86
16.265	49	44.211	71	59.279	61	14.343	58	48.327	89			14.013	1	**25.386**	**23**	38.359	14
17.277	38	45.157	93			15.312	60	49.317	83			21.198	13	26.381	42	43.107	18
18.235	44	46.283	86			16.251	64	50.317	69			22.222	14	27.368	29	45.136	166
19.243	44					17.262	60	54.040	124			**25.348**	**65**	29.440	46	47.156	16
20.393	50					18.232	70	55.253	102			28.106	2	31.010	12	48.138	15
21.350	58					19.241	68					42.023	14	31.530	47	56.029	179
22.342	36					20.392	78					53.211	95	32.464	42		
23.288	50					21.341	76							33.391	41		
24.407	32													34.383	45		
25.380	**50**													35.448	42		
26.380	33													36.390	40		
27.367	38													37.439	36		
28.446	44																
29.443	33																
		Total = 3250	Use 3250	Total = 2952		Total = 2952		Total =3768		Total = 1812		Total = 325		Total = 2130		Total = 806	
						58.032	379										

Notes — Peter Pan: No performances in '62 or 70; Est. 300 more in 60s & 70s; Followed by P. Pan on ice, pantomime, musical, etc. Use 3250.

Charley's Aunt: George Abbott's version: musical play, 2 acts, Where's Charley?

Doll'sHouse: Runs in 1921 are unknown, 13 is just a guess.

Knights..bold: Runs in 1936 are a guess.

Doctor's Dilemma 5/T, 3-5acts Lon.11/1906 BernardShaw		Treasure Island 4/Play, 4a Lon.12/1922 J. B. Fagan		Lilac Time 1,4/MusP,3a US, 3/1921 A. Ross FranzShubert		Saint Joan 2,3/Play,6a+ NY, 12/1923 Bernard Shaw		YellowSands 5/Comedy, 3a First run below E.&A.Phillpotts		Escape 5/episodes 1st run below J.Galsworthy		Tons of Money 1/Farse, 3a Eng. 3/1922 Evans & Valentine		Dick Whittington 4/Comedy,3a Footnote 1 Fnote 1again Harris, et al		Androcles & the Lion 4/Fable, 4a Eng. 9/1913 BernardShaw	
Serial	# Prfs	Serial	# Prfs	Serial	# Prfs	Serial	# Prfs	Serial	# Prfs	Serial	# Prfs	Serial	# Prfs	Serial	# Prfs	Serial	# Prfs
06.295	50	22.347	137	22.346	628	24.120	244	26.304	612	12.252	1	22.104	743	92.364	120	13.282	63
13.383	25	23.295	55	25.400	90	25.007	132	28.318	24	26.231	242	27.219	1	94.296	123	25.403	42
21.046	28	24.419	36	27.370	68	26.064	53	45.032	103	32.033	12	30.423	18	98.201	88	30.062	23
23.073	28	25.404	25	28.452	73	30.207	11	32.123	12	32.257	20	31.554	7	08.354	117	34.235	8
24.347	1	26.391	24	30.180	56	31.123	48	45.032	103			32.420	28	11.372	106	34.269	28
25.257	16	29.4495	27	32.465	35	34.342	35			Show 52.290				19.248	114	43.016	40
26.329	37	30.449	20	42.131	80	39.037	2			is entirely		See note 2.		23.294	94	55.056	20
30.012	8	31.549	27	49.042	149	47.313	82			different.				24.412	54	—	
39.034	35	33.395	23	Total = 1179		Total = 607		Total = 854		Total = 275		Total = 797		24.412	54	Total = 224	
39.075	95	34.386	22											25.399	122		
42.029	474	36.388	18											31.546	50		
56.213	68	37.443	35											32.466	84		
—		38.382	18											34.385	128		
Total = 865		46.286	37											38.003	10		
		47.329	37											49.318	86		
		48.325	52											52.299	105		
		49.324	46											56.281	94		
		50.324	20											—			
		—												Total = 1549			
		Sub total = 659															
		Goodman's version: 50.322	24														
		Littlewood: 53.345	36														
		56.297	31														
		Miles: 59.270	128														
		Total = 878															

Note 1. Some runs called "Dick Whittington & his Cat" Most revivals have the same characters, but major changes in the cast and/or management. Wearing marks several runs as first performance, evidently considering them different

Note 2. Run in '27 is unknown, but the company, London Repertory, performs many one-night stands.

Table E-3 (cont.)

Ghost Train 4,5/P,3acts Eng. 6/25 Arnold Ridley AR		Rosmersholm 5/Drama,3or4 Xlation: 2/1891 Beata's adapt Henrik Ibsen		Prisoner of Zenda 1/RP, 4a+ 1/1896 E.Rose adapt of A. Hope		Blue Bird Date 4/6a Xltn:12/09 Maeterlinck		Sport of Kings 3/Domes.C,3 First run below Ian Hay		White Cargo 2,3,4/P,3acts NY, 11/23 Leon Gordon		Month in the Country 5/C, 4acts 1st translated I. Turgenev		Way of the World 2/Comedy,5 England 1700 Wm.Congreve		Bohemian Girl 4/Opera,3 Lon. 1843	
Serial #	Prfs	Serial #	Prfs	Serial #	Prfs	Serial #	Prfs	Serial #	Prfs	Serial #	Prfs	Serial #	Prfs	Serial #	Prfs	Serial #	Prfs
25.351	655	91.040	2	96.005	255	10.331	81	24.294	319	24.167	821	26.209	16	24.061	158	1843	
27.107	36	92.103	22	97.172	9	11.373	70	26.086	16	27.059	32	26.276	30	27.324	96	:	
29.446	34	93.130	4	00.018	11	22.350	58	27.225	12	27.139	6	36.285	37	48.255	32	90.077	7
30.131	12	95.047	3	09.042	59	23.293	46	29.432	34	29.180	12	37.156	39	53.026	83	91.285	2
34.374	36	06.139	2	11.044	97	25.382	60	30.260	12	35.404	127	43.020	313	56.266	67	91.360	1
36.099	12	08.032	8	23.198	110					48.136	16	49.302	17			92.327	3
See note 3		12.078	2							all 3						93.061	7
—		12.189	1													93.306	1
Total = 785		17.141	4	Total = 541		Total = 315		Total = 393		Total = 1014		Total = 452		Total = 436		97.024	1
		26.268	52													97.239	2
		36.063	12													99.023	1
		48.142	17													20.304	4
		50.182	38													21.351	3
		59.252	124													22.358	22
		—														24.021	2
		Total = 291														24.398	3
																26.002	2
																31.203	
																31.372	
																51.171	
																—	
																Total unknown	

Note 3. Run in '27 from Apr. 30 to May 27.
Estimate 4 wks. @ 9 prfs/wk = 36.

Appendix F

Extinction rates of prehistoric taxa

Recall three ways you can verify that a process/entity obeys Gott's survival predictor:

- Enact the whole process of making a prediction and watch it play out. This is too slow—once is enough. Gott did that for his New York stage productions [11].
- Backing off one level, you can examine the statistics for the duration of a process and thereby confirm the prior probability Q, Equation 1. Then GSP follows as in Equation 2.
- Backing off one last step, you can show that extinction rate has uniform probability density. This leads to the prior and then to GSP via the argument in Appendix A.

The second is easiest. Chapter 2 treats the second method, and this appendix the third. It shows that extinction rates of certain prehistoric taxa are distributed uniformly if you are willing to make an assumption discussed below.

Leigh Van Valen has studied these rates and published extensive data [29]. His Table 1 lists extinction rates for 20 families, 38 genera, and 3 species. I used his data for genera simply because that sample is the biggest of the three.

Table F below shows these data rearranged in order of increasing extinction rate. The first column shows the rates in a curious unit. One macarthur equals $\ln 2/500$ years, the rate that gives a half-life of 500 years. But the unit is irrelevant here; we need only show that these rates are uniformly distributed. The second column in the table shows the number of occurrences in Van Valen's data. Clearly the least rates occur more often than the greatest, but this is probably a sampling bias. The longer a species survived, the more likely a scientist would discover its remains millions of years later and find them statistically significant for inclusion in a published table. Therefore, to adjust for those uncounted, I make a plausible but

Table F. Van Valen's data: extinction rates for prehistoric taxa.

Extinction rate in micro-macarthurs	Number of occurrences in Van Valen's list of genera	Product, which removes sampling bias	Cumulative sum for Figure F, which verifies uniform distribution
20	5	100	100
25	6	150	250
30	7	210	460
35	3	105	565
45	1	45	610
50	3	150	760
60	1	60	820
80	1	80	900
120	3	360	1,260
150	3	450	1,710
160	1	160	1,870
170	1	170	2,040
180	1	180	2,220
200	1	200	2,420
220	1	220	2,640

speculative assumption that the actual abundance was proportional to the occurrences multiplied by the hazard rate. Accordingly I multiply the second column by the first, which gives the adjusted numbers in the third column.

In Figure F an inset shows a histogram of these products. It is consistent with a uniform distribution, but statistical fluctuations mask that uniformity. A more effective way to suppress statistical noise is to run a cumulative sum, the total adjusted number of genera having an extinction rate less than each tabulated value. This appears in the fourth column of Table F. For uniform distribution this cumulative sum should increase linearly, and indeed it does as the main plot in Figure F shows. The data fit the straight line as well as you can expect for a sample of 38 data. It is amazing how all these critters know to die on schedule.

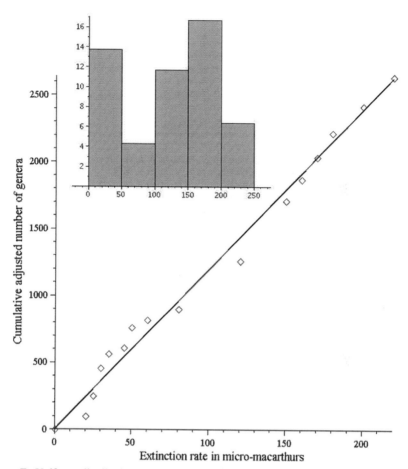

Figure F. Uniform distribution of extinction rates for prehistoric taxa Histogram shows number of genera in each interval, but statistical variations mask the uniformity. The cumulative sum of the numbers (main graph) averages out the fluctuations, and the resulting straight line makes a more convincing graphic.

Appendix G

Disaggregated mortality

In Equations 6 through 9 and beyond we have exponents that express the relative importance of hazards or classes of hazards. But "express" is too vague. Suppose we had abundant mortality data for things that have succumbed to multiple threats. Then how can we calculate these exponents? Let us prepare by reviewing a very simple case.

A big population P is exposed to two hazards with constant hazard rates λ and μ. During time increment dt the risk is $(\lambda + \mu)\,dt$, and so the number that expire is this risk times the number still surviving:

$$dP = -(\lambda + \mu)P\,dt \qquad (G\text{-}1)$$

Divide by P and integrate:

$$\ln P = -(\lambda + \mu)t + \ln P_0 \qquad (G\text{-}2)$$

$$\lambda + \mu = [\ln(P_0/P)]/t \qquad (G\text{-}3)$$

$$P = P_0 \exp[-(\lambda + \mu)t] \qquad (G\text{-}4)$$

By counting survivors we can determine only the sum $\lambda + \mu$, not λ and μ separately, so we need more information. Let us (virtually) sort dead bodies by cause of death. Let M (for mortality) denote the dead body count:

$$M = P_0 - P \qquad (G\text{-}5)$$

and by Equation G-1,

$$dM = -dP = (\lambda + \mu)P\,dt \qquad (G\text{-}6)$$

Disaggregating by cause of death gives

$$dM_\lambda = \lambda P\,dt; \qquad dM_\mu = \mu P\,dt \qquad (G\text{-}7)$$

Integrate this using Equation G-4:

$$M_\lambda(T) = \lambda P_0 \int_0^T \exp[-(\lambda + \mu)t] \qquad \text{(G-8)}$$

$$M_\lambda(T) = \frac{\lambda P_0}{\lambda + \mu}(1 - \exp[-(\lambda + \mu)T]) \qquad \text{(G-9)}$$

and similar expressions for M_μ. Note that $M_\mu + M_\lambda + P = P_0$, thus accounting for all of the original population.

If we use historical data for which the entire population has expired, then,

$$M_\lambda(\infty) = \frac{\lambda P_0}{\lambda + \mu}; \qquad M_\mu(\infty) = \frac{\mu P_0}{\lambda + \mu} \qquad \text{(G-10)}$$

and the ratio is

$$M_\lambda(\infty)/M_\mu(\infty) = \lambda/\mu \qquad \text{(G-11)}$$

No surprise here, the ratio of hazard rates is the same as the ratio of body counts. We know $\lambda + \mu$ from the overall mortality rate, and so λ and μ can now be separately determined, the desired result.

#   #   #

From here on we divide all populations and body counts by P_0 so that they represent fractions of the initial population. Now $Q = P/P_0$ denotes the surviving fraction, which we also interpret as survival probability.

For a Gott process we have Equation 6 in the main text instead of Equation G-4:

$$Q(t) = \left(\frac{J}{J + T}\right)^{1-q} \left(\frac{K}{K + Z(T)}\right)^q \qquad \text{(copy of Equation 6)}$$

Instead of Equation G-2,

$$\ln Q = (1 - q) \cdot [\ln J - \ln(J + T)] + q \cdot [\ln K - \ln(K + Z)] \qquad \text{(G-12)}$$

And instead of G-1, differentiate G-12 to get the dual hazard rates:

$$\frac{1}{Q}\frac{dM}{dT} = -\frac{1}{Q}\frac{dQ}{dT} = \frac{1-q}{J+T} + \frac{q}{K+Z}\frac{dZ}{dT} \qquad \text{(G-13)}$$

Like the transition from Equation G-6 to G-7 we can break out mortality rates for the separate hazards:

$$\frac{dM_j}{dT} = Q\frac{1-q}{J+T} \qquad \text{(G-14)}$$

$$\frac{dM_k}{dZ} = Q\frac{q}{K+Z} \qquad \text{(G-15)}$$

Unlike λ and μ the hazard rates (the fractions following Q) are not constant. They diminish slowly as frail entities die off leaving hardy survivors. (Recall that we are not dealing with things that wear out in time. Instead they develop survival skills.)

Integration gives the cumulative probability of succumbing to each hazard prior to time T. Using Equation 6 for Q, we find

$$M_j(T) = J^{1-q}K^q \int_0^T \frac{(1-q)\, dt}{(J+t)^{2-q}(K+Z(t))^q} \tag{G-16}$$

$$M_k(T) = J^{1-q}K^q \int_0^T \frac{q(dZ/dt)\, dt}{(J+t)^{1-q}(K+Z(t))^{1+q}} \tag{G-17}$$

In general one must integrate these equations numerically because $Z(t)$ is an arbitrary function. If we had historical mortality data like those in Sections 2.2, 2.3, and 3.3 (but disaggregated by cause of death), then we could try many different values of q and converge on the one that gives the best fit to observed mortality.

#   #   #

In one special case where $Z(T) = T$, we can do the integration analytically. This is the impresario's case discussed in Section 3.2. The two classes of hazards both depend on time but typically have different gestation periods. A brief look at this case offers further insight.

Integration of Equations G-16 and G-17 gives

$$M_j = \frac{K}{K-J}\left[1 - \left(\frac{1+T/K}{1+T/J}\right)^{1-q}\right] \tag{G-18}$$

$$M_k = \frac{J}{J-K}\left[1 - \left(\frac{1+T/J}{1+T/K}\right)^q\right] \tag{G-19}$$

One can show that $M_j + M_k + Q = 1$, thereby accounting for all the initial entities, the live ones and two categories of dead ones.

For historical data with no survivors,

$$M_j(\infty) = K^q \frac{J^{1-q} - K^{1-q}}{J - K} \tag{G-20}$$

$$M_k(\infty) = J^{1-q} \frac{J^q - K^q}{J - K} \tag{G-21}$$

and the ratio is

$$R = \frac{M_k(\infty)}{M_j(\infty)} = \frac{J^{1-q}}{K^q} \frac{J^q - K^q}{J^{1-q} - K^{1-q}} \tag{G-22}$$

In the limit $K \to J$,

$$R \to (1-q)/q \tag{G-23}$$

Just like the simple exponential case, the ratio of body counts gives the ratio of exponents. But this does not hold when the gestation times are unequal. In the limit $J \gg K$,

$$R = (J/K)^q \tag{G-24}$$

This says that the hazard with the shorter gestation takes the greater toll, especially when its exponent is big, because it strikes when the victim is still vulnerable before the frail ones have died out.

Appendix H

Stage productions with dual cum-risks

This appendix is a demonstration that a theoretical dual-risk formula works with real statistics, namely survival data for stage productions that first opened in London during the five years from 1920 through 1924.

The survival statistics of shows that expired long ago should approximate the formula $Q(S, T)$ given in Equation 9. Q is the probability of its survival beyond S performances that have played over a period of T days. A long-running show has small Q indicating that there are few survivors when it expires. Likewise a show that fails after a few performances has a large Q indicating that most productions in its cohort outlive it.

Equation 9 is a bivariate formula for Q having four parameters, J, K, q, and p. I added two more, a and g, that make a small correction for obsolescence, which makes a total of six as follows:

$$Q = \left(\frac{1}{1 + T/J}\right)^q \left(\frac{1}{1 + S/K + a(S/K)^g}\right)^p \tag{H}$$

Evaluation begins by assigning a rank to each show: rank $= 1$ for the longest running, rank $= 2$ for the next longest, and so forth up to one-night stands. Next, choose a set of parameters to be tested for accuracy. Using the formula, evaluate Q for all stage productions in the ensemble. If the formula is perfect and the ensemble infinitely big, then Q for each show will equal its fractional rank in that ensemble. The accuracy of the fit is simply the standard deviation of the calculated Qs from the set of ranks. Finally, test other trial sets of parameters in a sequence that converges on the optimum.

In Figure H-1 below the straight line represents the perfect formula and infinite ensemble. The univariate formula based only on performances gives the points with crosses, while the bivariate formula gives the improved fit shown as circles. With six parameters to play with, you would expect a better fit just from having additional adjustments available. However, the optimum parameter values are plausible in all

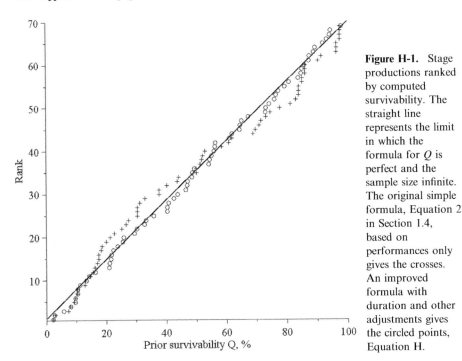

Figure H-1. Stage productions ranked by computed survivability. The straight line represents the limit in which the formula for Q is perfect and the sample size infinite. The original simple formula, Equation 2 in Section 1.4, based on performances only gives the crosses. An improved formula with duration and other adjustments gives the circled points, Equation H.

cases and consistent with earlier interpretations. In particular, both exponents p and q are always positive, their sum is close to 1.0, and the gestations K and J are consistent with plausible preparations and delays, all of which inspire confidence in the formulation.

Recall that the criterion for identifying the optimum set of parameters is its minimum *standard deviation* σ from the ideal straight line in Figure H-1. The search for this minimum has an interesting quirk. As you scan along some line or curve in parameter space, σ briefly varies smoothly until Q for one show passes Q for another, at which point they swap ranks in the list of Qs, which makes a small discontinuity. Many of these discontinuities produce jagged plots of σ as shown in Figure H-2. This particular scan is especially interesting because it demonstrates that the sum of exponents is very close to 1.0. As discussed in Section 3.1 and elsewhere, the sum must be 1.0 for fundamental reasons, and indeed this plot shows the minimum at about 1.012, which is well within the margin of error. Again, this builds confidence in the formulation, but nothing ever proves it rigorously.

Most computer programs for optimization expect continuous functions. I know of none that I would trust to correctly manage anything this jagged, nor have I any confidence in my ability to write one. Therefore, I found the minimum σ quasi-manually by using the computer only to make plots like Figure H-2 and then choosing an "eyeball" minimum for the next plot. Iterations converged nicely to a unique minimum.

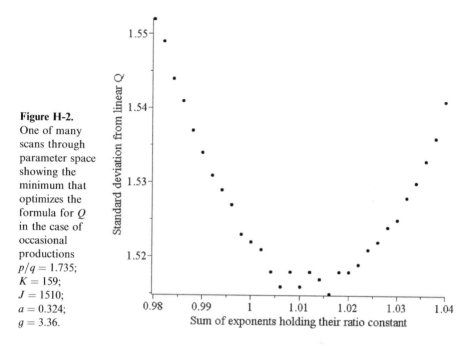

Figure H-2.
One of many
scans through
parameter space
showing the
minimum that
optimizes the
formula for Q
in the case of
occasional
productions
$p/q = 1.735$;
$K = 159$;
$J = 1510$;
$a = 0.324$;
$g = 3.36$.

For the occasional productions Figure H-3 shows a scan along the J-axis, which is particularly jagged. However, one can still pick a reasonable midpoint in the cluster of points at the bottom and call it the effective minimum. This optimum occurs at about four years, which is longer than I expected. However, on second thought this may be about the time a typical production in this class faces recurring costs for a second revival and a tough decision as to whether the risks are worth taking.

The J-axis scan for shows in the *main sequence* appears in Figure H-4. It has by far the most poorly defined minimum in this investigation. We can reasonably expect *some* quantity to be poorly defined for this ensemble because it does not have a good spread in the T, S plane, Figure 17. Instead, all points lie almost in line on that scatter diagram. And yet the curve does have a minimum, and it works quite well to produce the plausible results that appear in Section 3.3. For example, the statistical weight of duration is much stronger for occasional shows, which are more vulnerable to loss of personnel and public interest.

Finally, Tables H-1 and H-2 list the raw data from Wearing's book *The London Stage, 1920 to 1929*. Let us define *occasional productions* as those that had fewer than two performances per week over the long-term average. These appear in Table H-1. For example, the first entry *Gallant Cassian* opened in 1920 with serial number 20.163, column 10. It played only once more in 1928 with serial number 28.064. The ratio of days elapsed to number of performances was 1,421, column 9, which is just about as infrequent as one ever finds.

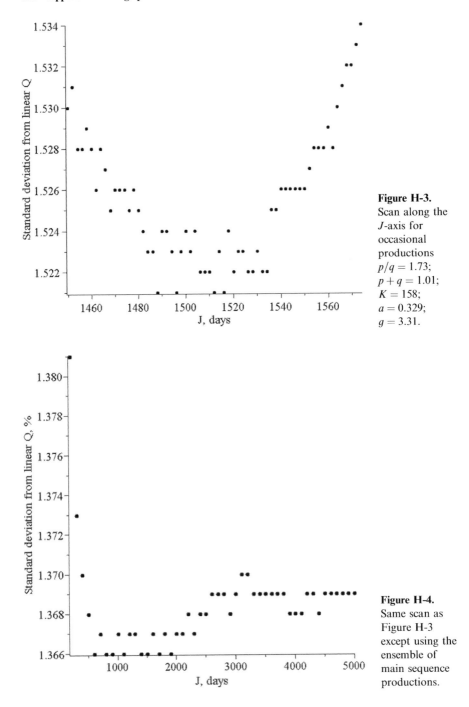

Figure H-3.
Scan along the
J-axis for
occasional
productions
$p/q = 1.73$;
$p + q = 1.01$;
$K = 158$;
$a = 0.329$;
$g = 3.31$.

Figure H-4.
Same scan as
Figure H-3
except using the
ensemble of
main sequence
productions.

The shows called *main sequence* were performed more frequently. They are listed in Table H-2. Typically they played eight times per week, six nights and two matinees. For both tables the entries are listed in order of increasing frequency, which means decreasing ratio of days to performances. Both tables have a column of "Links", which gives Wearing's serial number for other shows performed the same day in the same theater and included in the price of admission. For example, in the fourth row of Table H-1, *From Life* appeared as 24.049 linked with a group of six short shows numbered 20.047 through 20.054. Then the show reappeared on its own in 1928. Had it not reappeared solo, I would not have listed it because its survival would presumably depend on the popularity of the other shows in the group.

"Rank" in column 5 of both tables refers to the production's survival rank beginning with 1 for the show that survived the most cum-risk and ending with 316 for the one that succumbed to the least.

Table H-1. Survival data for occasional stage productions that opened in London from 1920 through 1924

Short Name	Links	Total Perfs	Rank	Date Opened	Date Closed	Dura-tion, days	Ratio	First Serial No.	Ps		Serial	Ps		Revivals, Notes Serial	Ps	Notes	Serial	Ps
Gallant Cassian		2	310	18-May-20	26-Feb-28	2841	1420.50	20.163	1	#	28.064	1						
His Majesty's P…		2	295	11-Feb-21	15-Feb-25	1466	733.00	21.036	1	#	25.051	1						
Antony Settles Down	none	2	301	17-Dec-22	31-May-26	1262	631.00	22.333	1	#	26.144	1						
From Life	47…54	3	293	3-Feb-24	15-Apr-28	1534	511.33	24.049	2	=	28.141	1						
Hugh the Drover		14	242	14-Jul-24	13-Apr-39	5387	384.79	24.275	3	#	37.117	5	=	38.010	3	=	39.102	3
Snowflakes	112..3	13	253	12-Apr-20	24-Sep-27	2722	209.38	20.113	6	=	24.334	2	=	25.311	1	=	27.257	4
Happy New Year	166..70	30	196	31-May-22	5-Mar-39	6123	204.10	22.166	29	#	39.050	1						
Banana Girl	47…54	3	294	3-Feb-24	27-Sep-25	603	201.00	24.054	2	=	25.271	1						
David Garrick (S)		29	206	9-Dec-20	9-Jun-32	4201	144.86	20.377	1	#	22.058	27	#	32.208	1			
Flora's Awakening		12	260	26-Apr-20	19-Sep-24	1608	134.00	20.130	10	#	24.314	2						
Pleasure Garden		33	191	29-Jun-24	19-Oct-35	4130	125.15	24.255	2	#	29.071	17	#	35.348	14			
Defeat, The	72..3	7	279	14-Mar-20	22-Apr-22	770	110.00	20.073	1	\	22.109	6						
Amarilla	112..3	27	212	12-Apr-20	22-Sep-27	2720	100.74	20.112	16	=	24.299	6	=	25.276	3	=	27.256	2
Yetta Polowski		12	257	12-Nov-22	11-Apr-25	882	73.50	22.302	1	#	25.089	11						
Pick Me Up	306…11	3	291	14-Sep-24	20-Apr-25	219	73.00	24.311	1	=	25.101	2						
Progress (Ervine)	92..4	67	146	3-Apr-22	9-Apr-35	4755	70.97	22.093	63	#	35.130	3	#	Omit 37.314, unrelated.				LGG
Sybarite, The	151..2	8	275	5-May-24	19-Sep-25	503	62.88	24.152	1	#	25.250	7						
Amends	19..23	73	135	25-Jan-22	5-Aug-32	3846	52.68	22.019	72	#	32.267	1						
Family Man, A		108	113	2-Jun-21	15-Jun-35	5127	47.47	21.152	51	#	27.244	37	#	35.188	20			
Taffy		25	217	26-Feb-23	10-Oct-25	958	38.32	23.047	1	#	25.258	24						
Windows		94	121	25-Apr-22	6-Feb-32	3575	38.03	22.116	39	#	27.207	21	=	32.009	34			
New Morality		45	173	28-Nov-20	1-Aug-25	1708	37.96	20.353	2	#	21.145	3	#	25.203	40			
Eight O'Clock	384..7	120	98	15-Dec-20	12-Aug-32	4259	35.49	20.384	107	#	32.227	7	#	32.272	6	Guess 7 perf.		

Short Name	Links	Total Perfs	Rank	Date Opened	Date Closed	Dura-tion, days	Ratio	First Serial No.	Ps		Serial	Ps		Serial	Ps		Serial	Ps
Doll's Romance	none	14	244	9-Mar-20	24-Jun-21	473	33.79	20.066	3	=	20.099	9	#	21.176	2			
Rumour		69	144	3-Dec-22	20-Apr-29	2331	33.78	22.321	2	#	29.055	67						
Luck King		2	305	16-Oct-21	16-Dec-21	62	31.00	21.256	1	#	21.353	1						
Vortex, The		339	28	25-Nov-24	17-May-52	10036	29.60	24.385	244	#	29.064	12	#	52.041	83			
Cupboard Love	19..23	73	137	25-Jan-22	7-Nov-27	2113	28.95	22.023	72	#	27.311	1						
Dead Man's Pool	80..85	149	84	28-Mar-21	12-Sep-32	4187	28.10	21.083	104	=	24.220	38	#	32.292	7		Footnote	
Aladdin & ... Lamp		234	57	21-Dec-20	22-Jan-38	6242	26.68	20.399	185	#	37.437	49						
Circle, The		397	19	3-Mar-21	19-Mar-49	10244	25.80	21.058	180	#	More, in Table of Many Revivals.							
Lure, The		65	151	25-Mar-23	2-Sep-27	1623	24.97	23.067	1	#	24.160	52	#	27.235	12			
E. & OE.	249..53	163	76	12-Oct-21	5-Aug-32	3951	24.24	21.253	119	#	24.219	38	#	32.269	6		LGG	
Arent We All		206	65	10-Apr-23	18-Mar-35	4361	21.17	23.078	110	#	29.078	92	#	35.104	4			
Wrong No. (O & F)		186	69	16-Jun-21	4-Sep-31	3733	20.07	21.169	162	#	29.080	12	#	31.349	12		Footnote	
Our Ostriches		120	97	14-Nov-23	31-May-30	2391	19.93	23.262	91	=	30.164	29						
In the Zone	164..7	21	226	15-Jun-21	22-Jul-22	403	19.19	21.166	14	=	22.206	7						
Comedy..Good & Ev		15	237	6-Jul-24	11-Apr-25	280	18.67	24.263	1	#	25.084	14						
If		193	67	30-May-21	22-Feb-31	3556	18.42	21.148	180	#	31.048	13						
Tiger Cats		145	86	26-Jun-24	13-Jun-31	2544	17.54	24.251	7	=	24.280	116	=	31.230	22			
Mary Rose		568	7	22-Apr-20	22-Mar-47	9831	17.31	20.129	399	=	26.007	90	=	29.159	65	#	47.053	14
Eye of Siva		92	122	8-Aug-23	16-Dec-27	1592	17.30	23.193	86	#	27.349	6						
Outward Bound		359	25	17-Sep-23	7-Sep-40	6201	17.27	23.218	15	=	More, in Table of Many Revivals.							
Me & my Diary		68	145	19-Jan-22	17-Jan-25	1095	16.10	22.012	37	=	22.051	22	=	25.006	9			
Through the Crack		47	171	27-Dec-20	13-Jan-23	748	15.91	20.413	24	#	22.334	23						
Treasure Island		878	3	23-Dec-22	20-Feb-60	13574	15.46	22.347	137	#	More, in Table of Many Revivals.							
Way of the Eagle		162	77	20-Jun-22	19-Apr-29	2496	15.41	22.192	150	#	29.107	12						
Faithful Heart		250	52	16-Nov-21	6-Feb-32	3735	14.94	21.313	185	#	30.041	12	#	31.541	53			

Table H-1 (*cont.*)

Short Name	Links	Total Perfs	Rank	Date Opened	Date Closed	Dura-tion, days	Ratio	First Serial No.	Ps		Serial	Ps		Serial	Ps		Serial	Ps	Revivals, Notes
Grain of Mustard S.		288	38	20-Apr-20	6-Dec-30	3883	13.48	20.121	222	=	24.380	19	=	30.359	47				
Lonely Lady (Wife)		17	234	5-Jul-20	5-Feb-21	216	12.71	20.244	1	#	21.011	16							
Skin Game		465	14	21-Apr-20	5-May-35	5493	11.81	20.124	349	#	29.279	88	=	35.156	28				
Man with a Load of		293	37	7-Dec-24	21-Oct-33	3241	11.06	24.394	2	=	25.175	261	=	33.280	30				
In & Out		52	167	26-Apr-24	24-Oct-25	547	10.52	24.145	15	#	25.261	37							Revival called "Easy Money".
Plus Fours		163	81	17-Jan-23	23-Jul-27	1649	10.12	23.015	151	#	27.211	12							
It Happened in Ardo..		14	248	19-Oct-24	7-Mar-25	140	10.00	24.352	1	#	25.061	13							
Lillies of the Field		307	36	5-Jun-23	6-Jun-31	2924	9.52	23.135	269	#	31.190	38							
At Mrs. Beam's		315	32	27-Feb-21	14-Dec-28	2848	9.04	21.053	2	#	23.039	21	=	23.075	280	#	28.419	12	
Daily Bread	106..8	13	251	17-Apr-22	22-Jul-22	97	7.46	22.108	6	=	22.205	7							
Bulldog Drummond		497	9	29-Mar-21	1-May-31	3686	7.42	21.086	430	=	22.351	53	#	26.333	12	#	31.169	2	
Pelican, The		327	30	20-Oct-24	11-Apr-31	2365	7.23	24.353	243	#	28.408	12	#	31.042	72				
Loyalties		543	8	8-Mar-22	24-Sep-32	3854	7.10	22.063	407	#	28.298	96	#	32.281	40				
Secrets		464	15	7-Sep-22	18-Jul-31	3237	6.98	22.231	373	=	29.301	79	#	31.319	12				
Bill of Divorcement		484	11	14-Mar-21	4-Apr-30	3309	6.84	21.071	402	#	29.247	70	#	30.104	12				
Lavender Ladies		157	79	25-Feb-23	12-Dec-25	1022	6.51	23.044	1	#	25.222	156							
Woman & th Apple		32	194	12-Nov-20	28-May-21	198	6.19	20.328	28	=	21.130	4							More, in Table of Many Revivals.
Sport of Kings		393	20	8-Sep-24	18-Jul-30	2140	5.45	24.294	319	=	25.217	46	#	27.227	12				
Fake, The		268	44	13-Mar-24	26-Aug-27	1262	4.71	24.102	210	=									
Britain's Daughter		12	255	20-Nov-22	13-Jan-23	55	4.58	22.313	12										
Young Person...Pink		307	35	10-Feb-20	6-Oct-23	1335	4.35	20.033	1	#	20.095	208	#	23.185	98				
Paddy th Next Best..		976	2	5-Apr-20	2-Jan-31	3925	4.02	20.104	886	\	23.299	57	\	29.450	18	#	30.451	15	

For 1923..24, shows that first played outside London are not all listed.

Notes and legend on next page.

Omissions:

Wandering Jew.	Wearing's index erroneously lists 37.386.
	Performance total in 40.110 is unknown; guess 80.
ShallWeJoin...Ladies?	Six single performances are lumped together.
Escape	Show 52.290 is something entirely different.

Footnotes:

Wrong Number	New title in '31: "By Candle Light"
Dead Man's Pool	LGG, performed continuously from 2pm till midnight.
	Run in 1932 is unknown, guess 7.

Symbols between revivals:

=	Same production
#	Same composition, different production
‖	Major changes
\	Ambiguous

Table H-2. Table of main-sequence stage productions that opened in London from 1920 to 1924

Short Name	Links	Total Perfs	Rank	Date Opened	Date Closed	Dura-tion, days	Ratio	First Serial No.	Ps		Serial	Ps	Revivals, Notes Serial	Ps	Serial	Ps	Notes
Panto. that Petered	47..54	2	311	3-Feb-24	8-Feb-24	6	3.00	24.047	2								
Magdalen's Husband		6	281	1-Jan-24	17-Jan-24	17	2.83	24.001	6								
Co-Optimists		1812	1	27-Jun-21	11-May-35	5067	2.80	21.180	498								More, in Table of Many Revivals.
Wat Tyler		5	285	14-Nov-21	26-Nov-21	13	2.60	21.302	5								
Rounding...Triangle	182..6	180	70	29-Jun-21	16-Sep-22	445	2.47	21.184	117	#	22.211	63					
Fledglings		6	193	1-Nov-23	14-Nov-23	14	2.33	23.251	6	=	30.236	26					No relation
Rainbow, The		152	82	3-Apr-23	8-Mar-24	341	2.24	23.076	117	#	24.071	35					
Arthur		10	264	12-Mar-23	31-Mar-23	20	2.00	23.059	10								
Melloney Holtspur		8	273	10-Jul-23	25-Jul-23	16	2.00	23.183	8								
Human Touch		2	306	14-Mar-21	17-Mar-21	4	2.00	21.067	2								
Enthusiast, The		5	286	4-Jul-23	12-Jul-23	9	1.80	23.178	5								
Troth, The		6	280	2-Jul-23	11-Jul-23	10	1.67	23.175	6								
His Happy Home		8	274	5-Jan-20	16-Jan-20	12	1.50	20.002	8								
Life's a Game		6	283	18-May-22	26-May-22	9	1.50	22.152	6								
Our Nell		148	85	16-Apr-24	8-Nov-24	207	1.40	24.137	140	#	24.368	8					
Cheezo		24	219	15-Nov-21	17-Dec-21	33	1.38	21.308	24								
David Copperfield		30	202	6-Jun-23	14-Jul-23	39	1.30	23.136	30								
Nursery, The	74..6	4	289	15-Mar-20	19-Mar-20	5	1.25	20.074	4								
Havoc		175	72	4-Nov-23	31-May-24	210	1.20	23.256	1	=	24.020	174					
Faust on Toast		34	190	19-Apr-21	28-May-21	40	1.18	21.100	34								
Mask & the Face		273	42	5-Feb-24	13-Dec-24	313	1.15	24.059	41	=	24.185	232					
Teddy Tail		24	220	27-Dec-20	22-Jan-21	27	1.13	20.406	24								
Goddess		62	156	6-Jun-22	10-Aug-22	66	1.06	22.175	2	=	22.188	60					
Six-Cylinder Love		24	221	24-Dec-24	17-Jan-25	25	1.04	24.413	24								
At the Villa Rose		127	91	10-Jul-20	13-Nov-20	127	1.00	20.248	127	=	[20.397]	[23]					

Short Name	Links	Total Perfs	Rank	Date Opened	Date Closed	Duration, days	Ratio	First Serial No.	Ps	Revivals, Notes							
										Serial	Ps	Serial	Ps	Serial	Ps	Serial	Ps
How to be Happy...		25	216	1-Sep-20	25-Sep-20	25	1.00	20.272	25								
Pedlar's Pie		15	235	2-Jun-23	16-Jun-23	15	1.00	23.128	15								
Bushido	164..7	14	243	15-Jun-21	28-Jun-21	14	1.00	21.164	14								
Timothy		7	276	1-Oct-21	7-Oct-21	7	1.00	21.239	7								
Tyranny of Home		7	278	16-Dec-24	22-Dec-24	7	1.00	24.405	7								
Bargain	106..8	6	282	17-Apr-22	22-Apr-22	6	1.00	22.106	6								
Prince & the Groom		5	287	5-Apr-20	9-Apr-20	5	1.00	20.102	5								
Tartan Peril		4	290	25-May-21	28-May-21	4	1.00	21.129	4								
First & Last		3	292	30-May-21	1-Jun-21	3	1.00	21.144	3								
Three Days		2	296	27-Jan-24	28-Jan-24	2	1.00	24.036	2								
Progress (Munro)		2	297	20-Jan-24	21-Jan-24	2	1.00	24.024	2								
South Wind		2	298	29-Apr-23	30-Apr-23	2	1.00	23.090	2								
Three		2	299	20-Mar-21	21-Mar-21	2	1.00	21.073	2								
Kingd'm, Pwr, Glory		2	300	16-Jan-21	17-Jan-21	2	1.00	21.005	2								
Machine Wreckers		2	302	6-May-23	7-May-23	2	1.00	23.095	2								
Mental Athletes		2	303	18-Feb-23	19-Feb-23	2	1.00	23.036	2								
Parish Watchman		2	304	10-Jul-21	11-Jul-21	2	1.00	21.197	2								
Children's Carnival		2	307	20-Jun-20	21-Jun-20	2	1.00	20.224	2								
Forerunners		2	308	19-Dec-20	20-Dec-20	2	1.00	20.390	2								
There Remains ...	13..14	2	309	18-Jan-20	19-Jan-20	2	1.00	20.013	2								
Thank You, Phillips!		76	133	10-Nov-21	21-Jan-22	73	0.96	21.294	76								
First Kiss		43	175	10-Nov-24	20-Dec-24	41	0.95	24.374	43								
Mayfair & Montmart.		77	128	9-Mar-22	20-May-22	73	0.95	22.065	77								
Pomp & Circumst.		18	232	8-Jun-22	24-Jun-22	17	0.94	22.180	18								
After Dinner		35	189	12-Jul-21	13-Aug-21	33	0.94	21.202	35								

Symbols between revivals:

= Same production

\# Same composition, different production

‖ Major changes

\ Ambiguous

Table H-2 (*cont.*)

Short Name	Links	Total Perfs	Rank	Date Opened	Date Closed	Dura-tion, days	Ratio	First Serial No.	Ps	Revivals, Notes Serial	Ps	Serial	Ps	Serial	Ps
Yoicks !		266	46	11-Jun-24	14-Feb-25	249	0.94	24.229	266						
Tancred		14	240	16-Jul-23	28-Jul-23	13	0.93	23.186	14						
Alternative, The		14	241	12-Mar-23	24-Mar-23	13	0.93	23.058	14						
Clogs to Clogs		14	245	10-Nov-24	22-Nov-24	13	0.93	24.373	14						
G.H.Q. Love	273..5	110	109	1-Sep-20	11-Dec-20	102	0.93	20.273	110	London's Grand Guignol					
Far above Rubies		41	177	27-Mar-24	3-May-24	38	0.93	24.122	41						
Running Water		27	210	5-Apr-22	29-Apr-22	25	0.93	22.097	27						
What did Husband...		120	99	27-Sep-20	15-Jan-21	111	0.93	20.290	120						
Her Daughter		13	252	24-Jun-24	5-Jul-24	12	0.92	24.249	13						
Low Tide		13	254	12-Aug-24	23-Aug-24	12	0.92	24.281	13						
Medium (Thoma)		12	256	10-Jan-23	20-Jan-23	11	0.92	23.009	12	No relation to 48.078, Menotti					
Second Round		11	262	8-Nov-23	17-Nov-23	10	0.91	23.259	11						
Secret Agent		11	263	2-Nov-22	11-Nov-22	10	0.91	22.293	11						
Cabaret Girl		362	24	19-Sep-22	11-Aug-23	327	0.90	22.238	362						
London, Paris, & NY		366	23	4-Sep-20	30-Jul-21	330	0.90	20.278	366						
Orange Blossom		50	170	4-Dec-24	17-Jan-25	45	0.90	24.393	50						
Eileen		40	178	27-May-22	1-Jul-22	36	0.90	22.158	40						
Society Ltd.		20	229	24-Mar-20	10-Apr-20	18	0.90	20.085	20						
Tropic Line		10	265	23-May-24	31-May-24	9	0.90	24.177	10						
Monica		10	266	4-Apr-24	12-Apr-24	9	0.90	24.130	10						
Forest, The		58	160	6-Mar-24	26-Apr-24	52	0.90	24.090	58						
Husbands are Prob.		19	230	3-Aug-22	19-Aug-22	17	0.89	22.213	19						
Fairy Tale		28	207	6-Feb-24	1-Mar-24	25	0.89	24.060	28						
Primrose		255	51	11-Sep-24	25-Apr-25	227	0.89	24.304	255						
Changing Guard	19..23	72	140	25-Jan-22	29-Mar-22	64	0.89	22.020	72						

Short Name	Links	Total Perfs	Rank	Date Opened	Date Closed	Duration, days	Ratio	First Serial No.	Ps	Serial	Ps	Serial	Ps	Serial	Ps
										Revivals, Notes					
Wonderful Visit		27	209	10-Feb-21	5-Mar-21	24	0.89	21.033	27						
Love?!		18	231	4-Mar-21	19-Mar-21	16	0.89	21.061	18						
Royal Visitor		9	267	27-Sep-24	4-Oct-24	8	0.89	24.329	9						
Conchita		9	268	19-Mar-24	26-Mar-24	8	0.89	24.112	9						
Pilgrim of Eternity		9	270	12-Nov-21	19-Nov-21	8	0.89	21.298	9						
Fantasia (Put&Take)		9	271	21-Nov-21	28-Nov-21	8	0.89	21.318	9 =	21.331	[20]				
Man in Mary's Room	384..7	107	115	15-Dec-20	19-Mar-21	95	0.89	20.385	107						
MadamePompadour		461	16	20-Dec-23	31-Jan-25	409	0.89	23.289	461						
Old English		97	119	21-Oct-24	14-Jan-25	86	0.89	24.356	97						
Safety Match		229	59	13-Jan-21	3-Aug-21	203	0.89	21.003	229						
Prude's Fall		226	60	1-Sep-20	19-Mar-21	200	0.88	20.276	226						
Unknown, The		78	127	9-Aug-20	16-Oct-20	69	0.88	20.265	78						
Night Out		310	33	18-Sep-20	18-Jun-21	274	0.88	20.285	310						
Scandal		77	129	19-Sep-22	25-Nov-22	68	0.88	22.237	77						
To Have the Honour		196	66	22-Apr-24	11-Oct-24	173	0.88	24.142	196						
Just Like Judy		17	233	11-Feb-20	25-Feb-20	15	0.88	20.036	17						
Blue Lagoon		263	48	28-Aug-20	16-Apr-21	232	0.88	20.270	263						
Pansy'sArabianNts.		25	215	16-Aug-24	6-Sep-24	22	0.88	24.283	25						
Some Detective		25	218	16-Jul-21	6-Aug-21	22	0.88	21.207	25						
Hassan, and How...		282	40	20-Sep-23	24-May-24	248	0.88	23.222	282						
Storm...Tinderley		116	104	13-Aug-24	22-Nov-24	102	0.88	24.282	116						
Little Dutch Girl		215	62	1-Dec-20	7-Jun-21	189	0.88	20.361	215						
In the Snare		66	150	4-Jul-24	30-Aug-24	58	0.88	24.261	66						
SecondLittleRev..at9		173	73	18-Mar-24	16-Aug-24	152	0.88	24.109	173						
Zozo		74	134	4-Aug-22	7-Oct-22	65	0.88	22.214	74						

For 1923 and 24, shows that first played outside London are not all listed.

Table H-2 (*cont.*)

Short Name	Links	Total Perfs	Rank	Date Opened	Date Closed	Duration, days	Ratio	First Serial No.	Ps	Revivals, Notes					
										Serial	Ps	Serial	Ps	Serial	Ps
MidsummmMadness		115	105	3-Jul-24	11-Oct-24	101	0.88	24.259	115						
Green Cord		122	94	2-Jun-22	16-Sep-22	107	0.88	22.172	122						
Love among Paint...		73	136	30-Apr-21	2-Jul-21	64	0.88	21.111	73						
His Girl		81	126	1-Apr-22	10-Jun-22	71	0.88	22.088	81						
Head over Heels		113	106	8-Sep-23	15-Dec-23	99	0.88	23.205	113						
Rebel Maid		113	107	12-Mar-21	18-Jun-21	99	0.88	21.066	113						
Out to Win		121	95	11-Jun-21	24-Sep-21	106	0.88	21.161	121						
Three Graces		121	96	26-Jan-24	10-May-24	106	0.88	24.035	121						
League of Notions		359	26	17-Jan-21	26-Nov-21	314	0.87	21.008	359						
Fun of the Fayre		239	56	17-Oct-21	13-May-22	209	0.87	21.259	239						
Bluebeard's 8th Wife		482	12	26-Aug-22	20-Oct-23	421	0.87	22.224	482						
Beauty Prize		213	63	5-Sep-23	8-Mar-24	186	0.87	23.203	213						
Oh, Juliet		142	87	22-Jun-20	23-Oct-20	124	0.87	20.228	142						
Roof & Four Walls		134	90	16-Jan-23	12-May-23	117	0.87	23.014	134						
Good Luck		260	49	27-Sep-23	10-May-24	227	0.87	23.228	260			London's Grand Guignol			
Amelia's Suitors	92..4	63	155	3-Apr-22	27-May-22	55	0.87	22.092	63						
Gipsy Princess		212	64	26-May-21	26-Nov-21	185	0.87	21.132	212						
Latitude 15S	182..6	117	103	29-Jun-21	8-Oct-21	102	0.87	21.182	117						
Robert E. Lee		109	110	20-Jun-23	22-Sep-23	95	0.87	23.160	109						
Great Well		70	143	19-Dec-22	17-Feb-23	61	0.87	22.338	70						
Dancers, The		349	27	15-Feb-23	15-Dec-23	304	0.87	23.034	349						
No Man's Land		62	157	2-Dec-24	24-Jan-25	54	0.87	24.391	62						
Kate		31	195	25-Feb-24	22-Mar-24	27	0.87	24.082	31						
Dover St. to Dixie		108	111	31-May-23	1-Sep-23	94	0.87	23.126	108						
Money Doesn't Mattr		46	172	31-Jan-22	11-Mar-22	40	0.87	22.029	46						

Short Name	Links	Total Perfs	Rank	Date Opened	Date Closed	Duration, days	Ratio	First Serial No.	Ps	Revivals, Notes					
										Serial	Ps	Serial	Ps	Serial	Ps
M'Lady		23	222	18-Jul-21	6-Aug-21	20	0.87	21.208	23						
Whirled into Hap'nes	none	245	53	18-May-22	16-Dec-22	213	0.87	22.154	245						
Curate's Egg		84	124	22-Mar-22	2-Jun-22	73	0.87	22.078	84	22.146	[1]	One scene later called Bottles & Bones			
Cairo		267	45	15-Oct-21	3-Jan-22	232	0.87	21.255	267						
Cherry		76	132	22-Jul-20	25-Sep-20	66	0.87	20.263	76						
Mumsee		38	180	24-Feb-20	27-Mar-20	33	0.87	20.047	38						
You and I		38	181	30-Dec-24	31-Jan-25	33	0.87	24.422	38						
Love Habit		53	166	7-Feb-23	24-Mar-23	46	0.87	23.030	53						
Threads		30	198	23-Aug-21	17-Sep-21	26	0.87	21.221	30						
Will You Kiss Me?		30	199	16-Nov-20	11-Dec-20	26	0.87	20.331	30						
EveryWoman'sPriv.		30	200	28-Sep-20	23-Oct-20	26	0.87	20.293	30						
Beating on the Door		15	236	6-Nov-22	18-Nov-22	13	0.87	22.300	15	=					
My Nieces		172	74	19-Aug-21	14-Jan-22	149	0.87	21.220	172						
It's All Wrong		112	108	13-Dec-20	19-Mar-21	97	0.87	20.381	112						
Laughing Lady		164	75	17-Nov-22	7-Apr-23	142	0.87	22.309	164						
Charlot's Revue		216	61	23-Sep-24	28-Mar-25	187	0.87	24.327	216						
Carte Blanche		67	147	2-Mar-23	28-Apr-23	58	0.87	23.053	67						
Mr. Todd's Exprmnt		67	148	30-Jan-20	27-Mar-20	58	0.87	20.024	67						
Love's Awakening		37	182	19-Apr-22	20-May-22	32	0.86	22.113	37						
Little Revue...at 9		192	68	2-Oct-23	15-Mar-24	166	0.86	23.232	192	Unrelated to 24.109, "Second Little Revue at Nine"					
Claimant		44	174	11-Sep-24	18-Oct-24	38	0.86	24.303	44						
Toils of Yoshitomo		22	223	3-Oct-22	21-Oct-22	19	0.86	22.245	22						
Truth about Blayds		124	93	20-Dec-21	5-Apr-22	107	0.86	21.361	124						
Island King		160	78	10-Oct-22	24-Feb-23	138	0.86	22.251	160						
Voice Outside		58	161	28-Apr-23	16-Jun-23	50	0.86	23.089	58						

Table H-2 (*cont.*)

Short Name	Links	Total Perfs	Rank	Date Opened	Date Closed	Dura-tion, days	Ratio	First Serial No.	Ps	Serial	Ps	Serial	Ps	Serial	Ps	Serial	Ps
Sister's Tragedy	166..70	29	203	31-May-22	24-Jun-22	25	0.86	22.167	29								
Spanish Lovers		29	204	21-Jun-22	15-Jul-22	25	0.86	22.193	29								
Card Players		29	205	26-Apr-22	20-May-22	25	0.86	22.118	29								
Golden Moth		281	41	5-Oct-21	3-Jun-22	242	0.86	21.244	281								
Not in our Stars		72	139	4-Feb-24	5-Apr-24	62	0.86	24.058	72								
Shortest Story ..	006..7	72	141	17-Jan-21	19-Mar-21	62	0.86	21.006	72								
Nice Thing		36	185	12-Aug-20	11-Sep-20	31	0.86	20.266	36								
Hotel Mouse		36	187	6-Oct-21	5-Nov-21	31	0.86	21.246	36								
Social Convenience		71	142	22-Feb-21	23-Apr-21	61	0.86	21.051	71								
Decameron Nights		371	22	20-Apr-22	3-Mar-23	318	0.86	22.114	371								
Haricot Beans	249..53	119	102	12-Oct-21	21-Jan-22	102	0.86	21.249	119								
Other Peo's Worries		77	131	29-Mar-22	2-Jun-22	66	0.86	22.085	77								
Birds of a Feather		35	188	9-Apr-20	8-May-20	30	0.86	20.109	34	30.395	1						
Come In		28	208	1-May-24	24-May-24	24	0.86	24.149	28								
This Marriage		21	224	7-May-24	24-May-24	18	0.86	24.156	21								
Old House		21	225	23-Jun-20	10-Jul-20	18	0.86	20.229	21								
First Love		21	227	7-Apr-20	24-Apr-20	18	0.86	20.107	21								
Elopement, The		14	246	28-Aug-23	8-Sep-23	12	0.86	23.199	14								
Other Times		14	247	6-Apr-20	17-Apr-20	12	0.86	20.105	14								
Audacious MrSquire		14	249	19-Feb-24	1-Mar-24	12	0.86	24.074	14								
Reggie Reforms	112..3	7	277	2-May-21	7-May-21	6	0.86	21.112	7								
Ninth Earl		54	164	9-Mar-21	23-Apr-21	46	0.85	21.065	54								
Hour & Man (V&S)		27	211	11-Feb-21	5-Mar-21	23	0.85	21.039	27								
Punch Bowl		630	5	21-May-24	7-Nov-25	536	0.85	24.174	554	#	25.246	76					
Devil Dick		20	228	16-Nov-22	2-Dec-22	17	0.85	22.307	20								

Short Name	Links	Total Perfs	Rank	Date Opened	Date Closed	Duration, days	Ratio	First Serial No.	Ps	Revivals, Notes					
										Serial	Ps	Serial	Ps	Serial	Ps
Crime		65	152	28-Nov-21	21-Jan-22	55	0.85	21.323	65						
London Life		39	179	3-Jun-24	5-Jul-24	33	0.85	24.201	39						
Reckless Reggie		13	192	18-Jul-23	28-Jul-23	11	0.86	23.188	13						
Listening-In		26	214	5-Aug-22	26-Aug-22	22	0.85	22.215	26						
Crossing		13	250	29-Sep-20	9-Oct-20	11	0.85	20.294	13						
A to Z		428	17	11-Oct-21	7-Oct-22	362	0.85	21.247	428						
Will Shakespeare		62	159	17-Nov-21	7-Jan-22	52	0.84	21.315	62						
Wheel		139	88	1-Feb-22	27-May-22	116	0.83	22.030	139						
Matter of Fact		30	201	27-Apr-21	21-May-21	25	0.83	21.109	30						
Mr. Malatesta		12	258	30-Jun-21	9-Jul-21	10	0.83	21.187	12						
Sarah of Soho		12	259	23-Feb-22	4-Mar-22	10	0.83	22.052	12						
False Values		12	261	11-Sep-24	20-Sep-24	10	0.83	24.302	12						
Now & Then		77	130	17-Sep-21	19-Nov-21	64	0.83	21.232	77						
Right to Strike		82	125	28-Sep-20	4-Dec-20	68	0.83	20.292	82						
Enchanted Cottage		64	154	1-Mar-22	22-Apr-22	53	0.83	22.057	64						
Gaspers	80..85	104	116	28-Mar-21	21-Jun-21	86	0.83	21.080	104	London's Grand Guignol					
Odd Spot		107	114	30-Jul-24	25-Oct-24	88	0.82	24.279	107						
Pins & Needles!		241	55	13-May-21	26-Nov-21	198	0.82	21.121	241						
Romantic Age		127	92	18-Oct-20	29-Jan-21	104	0.82	20.305	127						
Battling Butler		243	54	8-Dec-22	23-Jun-23	198	0.81	22.326	243						
Jumble Sale		176	71	16-Dec-20	7-May-21	143	0.81	20.388	176						
Puppets!		255	50	2-Jan-24	26-Jul-24	207	0.81	24.002	255						
Rats!		264	47	21-Feb-23	22-Sep-23	214	0.81	23.041	264						
Nine O'clock Revue		385	21	25-Oct-22	1-Sep-23	312	0.81	22.282	385						
Pot Luck!		283	39	24-Dec-21	8-Aug-22	228	0.81	21.366	283						

Table H-2 (*cont.*)

Short Name	Links	Total Perfs	Rank	Date Opened	Date Closed	Duration, days	Ratio	First Serial No.	Ps	Revivals, Notes					
										Serial	Ps	Serial	Ps	Serial	Ps
Little Girl in Red		36	186	10-Dec-21	7-Jan-22	29	0.81	21.342	36						
Success		56	163	21-Jun-23	4-Aug-23	45	0.80	23.164	56						
Music Box Revue		120	100	15-May-23	18-Aug-23	96	0.80	23.104	120						
Thing that Matters		30	197	22-Dec-21	14-Jan-22	24	0.80	21.365	30						
Daisy		15	238	14-Sep-20	25-Sep-20	12	0.80	20.280	15						
Nuts in May		15	239	9-May-22	20-May-22	12	0.80	22.135	15						
Prodigal Daughter		5	284	17-May-22	20-May-22	4	0.80	22.151	5						
Looking Glass		57	162	2-Oct-24	15-Nov-24	45	0.79	24.338	57						
London Calling		318	31	4-Sep-23	10-May-24	250	0.79	23.202	318						
Just Fancy!		332	29	26-Mar-20	11-Dec-20	261	0.79	20.088	332						
Snap		233	58	11-Aug-22	9-Feb-23	183	0.79	22.219	233						
Night of Temptations		64	153	14-Apr-23	2-Jun-23	50	0.78	23.082	64						
You'd be Surprised		100	118	27-Jan-23	14-Apr-23	78	0.78	23.020	100						
Sunshine of th World		50	169	18-Feb-20	27-Mar-20	39	0.78	20.039	50						
Ring Up		136	89	3-Sep-21	17-Dec-21	106	0.78	21.226	136						
Puss! Puss!		155	80	14-May-21	10-Sep-21	120	0.77	21.122	155						
Yes!		119	101	29-Sep-23	29-Dec-23	92	0.77	23.229	119						
Jenny		66	149	10-Feb-22	1-Apr-22	51	0.77	22.036	66						
Isabel, Edward, &…		101	117	31-Mar-23	16-Jun-23	78	0.77	23.072	101						
What Money cn Buy		87	123	26-Sep-23	1-Dec-23	67	0.77	23.227	87						
Aladdin (Rolls)		72	138	26-Dec-21	18-Feb-22	55	0.76	21.374	72						
Savage & th Woman		151	83	3-Mar-21	25-Jun-21	115	0.76	21.059	151						
Naughty Princess		270	43	7-Oct-20	28-Apr-21	204	0.76	20.300	270						
Destruction		8	272	4-Dec-22	9-Dec-22	6	0.75	22.323	8						
Old Story	249..53	4	288	12-Oct-21	14-Oct-21	3	0.75	21.251	4						

Short Name	Links	Total Perfs	Rank	Date Opened	Date Closed	Duration days	Ratio	First Serial No.	Ps	rel	Serial	Ps	rel	Serial	Ps	Notes
Her Market Price		51	168	17-Apr-24	24-May-24	38	0.75	24.138	51							
Rose & the Ring		41	176	19-Dec-23	17-Jan-24	30	0.73	23.286	41							
Trump Card		53	165	10-Aug-21	14-Sep-21	36	0.68	21.218	53							
Queen of Chittore	209..10	9	269	24-Jul-22	29-Jul-22	6	0.67	22.209	9							
Boy of my Heart		62	158	6-Mar-20	14-Apr-20	40	0.65	20.061	62							
Whirl of the World		635	4	14-Mar-24	21-Mar-25	373	0.59	24.106	635							
Brighter London		603	6	28-Mar-23	15-Mar-24	354	0.59	23.069	603							
Rockets		491	10	25-Feb-22	9-Dec-22	288	0.59	22.053	491							
Leap Year		471	13	20-Mar-24	20-Dec-24	276	0.59	24.114	471							
Peep Show		421	18	14-Apr-21	15-Dec-21	246	0.58	21.096	421							
Jig-Saw!		307	34	16-Jun-20	11-Dec-20	179	0.58	20.218	307							
Chuckles of 1922		95	120	19-Jun-22	12-Aug-22	55	0.58	22.190	95							
Sinners		26	213	8-Nov-24	22-Nov-24	15	0.58	24.371	26							
Babes in th Wood(A)		108	112	27-Dec-20	26-Feb-21	62	0.57	20.407	108							
Radios		36	183	5-Mar-23	17-Mar-23	13	0.36	23.055	36							
Spangles		36	184	11-Dec-22	23-Dec-22	13	0.36	22.328	36							
Fisherman's Love		[13]		20-Feb-20	omit			20.041	1	=	20.065	3	~	20.100	9	Nothing to go on, omit
Proposal		[3]				1		20.1615	1	=	21.021	1	=	28.067	1	
Red Feathers	245..6		*			1		21.131	unk	=	21.174	7				
Punchinello		[1]	*	22-Jun-24				24.246	1	=	32.041	[4]	=	32.069	[20]	No relation
Joan of Memories	13..14	[2]	*	18-Jan-20	19-Jan-20	2		20.014	2							
Magic Wand	74..6	[4]	*	15-Mar-20	19-Mar-20	5		20.075	4							
Russian Folk Scenes	74..6	[4]	*	15-Mar-20	19-Mar-20	5		20.076	4							
Oh, Hell!!!	273..5	[110]	*			1		20.275	110							
Tragedy...Mr. Punch	384..7	[107]	*			1		20.387	107							LGG

Table H-2 (*cont.*)

Short Name	Links	Total Perfs	Rank	Date Opened	Date Closed	Dura-tion, days	Ratio	First Serial No.	Ps	Serial	Ps	Serial	Ps	Serial	Ps
													Revivals, Notes		
Person Unknown	006..7	[72]	*	17-Jan-21	19-Mar-21	62		21.007	72						
Kill, The	80..85	[104]	*			1		21.084	104			London's Grand Guignol			
Chemist, The	80..85	[104]	*			1		21.085	104			London's Grand Guignol			
Toast	112..3	[7]	*			1		21.113	7						
Jealous Barbouille	164..7	[14]	*	15-Jun-21	28-Jun-21	14		21.167	14						
Old Women	182..6	[117]	*			1		21.185	117						
Shepherd's Pie	182..6	[117]	*			1		21.186	117						
Unseen	249..53	[119]	*	12-Oct-21	21-Jan-22	102		21.250	119						
De Mortuis	19..23	[72]	*	25-Jan-22	29-Mar-22	64		22.021	72						
Regiment, The	19..23	[72]	*	25-Jan-22	29-Mar-22	64		22.022	72						
Nutcracker Suite	92..4	[63]	*			1		22.094	63						
To Be Continued	166..70	[29]	*			1		22.168	29						
Better Half	166..70	[29]	*			1		22.170	29						
Dreamer Awakened	209..10	[9]	*	24-Jul-22	29-Jul-22	6		22.210	9						
Twinkle...Little Star	47..54	[2]	*	3-Feb-24	8-Feb-24	6		24.048	2						
Where....Clowns?	47..54	[2]	*	3-Feb-24	8-Feb-24	6		24.050	2						
Great Belief	47..54	[2]	*	3-Feb-24	8-Feb-24	6		24.052	2						
Bill Burke	47..54	[2]	*	3-Feb-24	8-Feb-24	6		24.053	2						

* Each of these stage productions is represented by a different member of its linked group.

Appendix I

Overall plan for survivability calculation

The main text contains a logic diagram in two parts that describes the overall calculation of survivability. Part 1 is in the introduction to Chapter 2; Part 2 in Section 3.4. The following outline puts it all together with additional detail. The numbering here is *ad hoc* lacking any simple connection to chapter and section numbers in the main text.

1.0   Find the simplest survival formula for hazards that are constant in time, Equation 1.

    1.1   Theoretical methods:

        1.1.1   Best estimate when the hazard rate is unknown or random, Section 1.1 in text.

        1.1.2   Formula based on an observer's random arrival time (related to doomsday argument), Section 1.5.

        1.1.3   Formula based on Bayes' theorem, Appendix D.

    1.2   Substantiating statistics:

        1.2.1   Survival of business firms, Section 2.2 in text.

        1.2.2   Survival of stage productions, Section 2.3.

        1.2.3   Other clues:

                1.2.3.1   Extinction rates of prehistoric taxa, Appendix F.

                1.2.3.2   Relationship to Zipf's law, Section 2.4

2.0   The simplest formula is now well established. Generalize to hazards that vary with time, as do man-made high-tech hazards to human survival. Theory is based primarily on 1.1.1 (above), secondarily on 1.1.2. Survival depends on risk exposure (cum-risk) prior to observation.

3.0   Generalize again to include two independent risk rates, Chapter 3. Normally one is constant in time, which represents natural hazards. The second is accelerating, as are man-made hazards.

4.0   Substantiate both 2.0 and 3.0 (above) using London stage productions.

    4.1   Main cum-risk is number of performances, which depletes paying attendance.

    4.2   Secondary cum-risk is duration, which involves issues of steady employment for the team and/or startup costs for revivals.

5.0   Formulate survivability of the human race using 2.0 and 3.0 (above), and Chapter 4 in the text.

    5.1   For natural hazards cum-risk is time.

    5.2   Find a formula for cum-risk due to man-made hazards, Section 4.2 in the text.

        5.2.1   Formula for economic development

        5.2.2   Tables of world population, gross world product, various indicators of progress in hazardous technologies

        5.2.3   Statistics of development in hazardous technologies

6.0   Calculate survivability as it depends on population-time (pop-time).

    6.1   Survival of civilization, Section 4.4

    6.2   Survival of the human race, Section 4.5

    6.3   Calculate the probability that civilization's collapse will rescue the human race by taking out man-made threats.

Appendix J

Multiple hazards

In Equation A-1 the system's probability of survival for time t is $\exp(-\lambda t)$, where λ is the constant risk per unit time. In general, λ varies with time, in which case this becomes

$$Q(t \mid \lambda) = \exp\left(-\int_0^t \lambda(s)\, ds\right) \tag{J-1}$$

The integral is a cum-risk, which reverts to λt when λ is constant.

As an example of time-dependent risk, consider the survival of an old house designated as a historic landmark. Urban sprawl has recently surrounded it. Since this has increased the price of land, the curator is concerned that the city council may sell the property to a developer for much needed cash. She assumes that the incentive to sell at time t is $\lambda = r \cdot m(t)$, where r is a constant, and m is the market value of the property, which a friendly appraiser updates quarterly.

Now Equation J-1 becomes

$$Q(t \mid r) = \exp(-rM), \quad \text{where } M(t) = \int m(t)\, dt \tag{J-2}$$

This integral is a cum-risk, which the curator maintains in her numerical risk model. Whenever she calculates a GSP, she uses the appropriate values for past and future:

$$M_p = \int_{-T_p}^{0} m(t)\, dt \quad \text{and} \quad M_f = \int_0^{T_f} m(t)\, dt \tag{J-3}$$

Time zero here is the time of her observation, and the two integrals separate the overall cum-risk into past and future.

Unfortunately the curator cannot estimate r in Equation J-2 without getting inside the minds of politicians, who answer her questions with evasive platitudes. The best she can do is to assume a probability density $F(r)$, which gives an estimate analogous to Equation A-2:

$$Q_m = \int_0^\infty F(r) \exp(-rM)\, dr \tag{J-4}$$

For reasons that will soon be apparent, she uses

$$F(r) = r^{\mu - 1}, \qquad 0 < \mu < 1 \tag{J-5}$$

Normalization of this pdf is not important because it cancels out in the posterior probability as we have seen with the prior constant/T, Equation B-3. Putting Equation J-5 in J-4 gives the prior probability:

$$Q_m = \Gamma(\mu)/M^\mu \tag{J-6}$$

a fractional power like those in Equation 8, which explains Equation J-5.

Another threat to the historic house is public apathy, the relatively few people who urge their elected officials to preserve it. The curator estimates an apathy hazard rate as the reciprocal of $v(t)$, the number of visitors who sign the guest book each month. The corresponding cum-risk is

$$A = \int \frac{dt}{v(t)} \tag{J-7}$$

By analogy to Equations J-2 through J-6, J-7 leads to

$$Q_a = \Gamma(\alpha)/A^\alpha, \qquad 0 < \alpha < 1 \tag{J-8}$$

Natural hazards such as fire, flood and earthquake are statistically constant—as likely to occur one day as another. So their risk variable is time T, hence:

$$Q_t = \Gamma(\tau)/T^\tau, \qquad 0 < \tau < 1 \tag{J-9}$$

The overall prior probability of survival is the product $Q = Q_m Q_a Q_t$. Substituting this in Equation B-4 converts the prior into the generalized GSP that we need:

$$G(f\,|\,p) = \left(\frac{1}{1 + M_f/M_p} \right)^\mu \cdot \left(\frac{1}{1 + A_f/A_p} \right)^\alpha \cdot \left(\frac{1}{1 + T_f/T_p} \right)^\tau; \qquad \mu + \alpha + \tau = 1 \tag{J-10}$$

Compare this to Equation 8. The correspondence provides the same theoretical support for the multivariate case that Appendix A provides for Gott's original predictor, Equation 2.

The sizes of exponents μ, α, τ indicate the relative importance of the corresponding hazards. Chances are that the curator does not know their values. Then the rule is to average over them while maintaining their sum = 1.0. This should be a weighted average governed by common sense. For example, $\tau = 0$ means no natural hazard. Since this never happens, it gets zero statistical weight in the average.

To conclude our story, the curator publishes an appeal with her gloomy GSP analysis, and shortly thereafter the council votes to auction the property. Our heroine chains herself on the front porch to a pillar supporting the second story. Meanwhile, a wealthy philanthropist reads her appeal. He rushes to the auction and makes the winning bid in the last second before the final gavel. Of course he falls in love with the curator. They marry and live happily for a few years until they realize that GSP for year 7 of their marriage has dropped below 30% confidence, and so they divorce and live happily until their next marriages.

Appendix K

Cum-risks for man-made hazards

The GSP for human survivability is given by the equation $G(f \mid p) = G_n^{1-q} \times G_m^q$ from Section 4.1. It is the geometric average of two quantities: G_n for natural threats and G_m for man-made. From Equation 10 the latter is explicitly

$$G_m = \frac{Z_p}{Z_p + Z_f} \tag{K-1}$$

The main task here is to find a mathematical formula for cum-risk Z, and then divide it into past and future to plug into this equation.

World population p is the pool of potential perpetrators, whether extinction is the work of a single mad scientist or the result of everybody's collective bad habits. So the hazard rate dZ/dt is proportional to the product of p by some measure U of the power and expertise humans have to commit the ultimate crime, whether it be deliberate or accidental. In calculus notation the equation $\Delta Z = p \times U$ from Section 4.1 becomes

$$dZ = U \cdot p \, dt \tag{K-2}$$

We need not worry about the units for measuring Z or any constant multiplier, because they cancel out in the ratio in Equation K-1. We shall express U in terms of population-time, which is defined as

$$X(t) = \int_{-\infty}^{t} p(y) \, dy \qquad \text{hence } dX = p \, dt \tag{K-3}$$

and Equation K-2 becomes

$$dZ = U \, dX \tag{K-4}$$

According to Equation 11, the equation for hazardous technology U is

$$\frac{dU}{dt} = C \cdot p \cdot U^{\mu}; \qquad \text{hence } U^{-\mu} \, dU = C \, dX \tag{K-5}$$

where C is a constant; population p is the pool of innovators; and U^μ is positive feedback from existing technology, which leverages further progress.

Let $M(t)$ denote all people-years lived after some date T_0 at which the technological feedback began, which turns out to be 1530 AD:

$$M(t) = X(t) - X(T_0); \qquad dM = dX \qquad \text{(K-6)}$$

Using Equation K-6 in K-5 and repeating K-4 gives the following fundamental equations to be solved:

$$dZ = U\, dM; \qquad U^{-\mu}\, dU = C\, dM \qquad \text{(K-7)}$$

There are five different solutions depending on the value of μ. All of them have elementary solutions listed in Table K below along with remarks on their behavior. To verify them, differentiate Z and U with respect to M and show that the results satisfy Equations K-7. During this process discard various annoying constant multipliers because they drop out anyhow in the ratio in Equation K-1. In the solution of Equations K-7, the constants of integration should make $U(0) = 0$ and $Z(0) = 0$. This holds in all cases when $\mu < 1$, but for $\mu \geq 1$, the technology U needs a seed to get started. For this same range, $\mu \geq 1$, there is some parameter (B or L) that needs to be evaluated, which reflects the unknown size of the seed. I do not know a systematic way to get this except by curve fitting when some technology spurts rapidly enough to call attention to the problem. In any case $\mu \geq 1$ is quite rare.

The quantity ω, which appears in the table, is given in Equation 12 from Section 4.2 copied as follows:

$$\omega = \frac{1}{1 - \mu} \qquad \text{hence } \mu = 1 - \frac{1}{\omega} \qquad \text{(K-8)}$$

Let us discuss four of the five solutions (rows) in Table K one at a time. The main case treated in the text, $\mu = 0.38$ and $\omega = 1.6$, falls in the first feedback range. It applies to the majority of hazards:

$$U = (\omega + 1)M^\omega; \qquad Z = M^{\omega+1} \qquad \text{(K-9)}$$

The next solution, $\mu = 1$, is exponential growth. A classic example is Moore's law for the growth of computer processing power, which doubles every few years. Well, not quite exponential, because it used to double every year.

The next solution, $1 < \mu < 2$, is scary because a few remote hazards could extend into this super-exponential range, and this solution has a finite pop-time at which $U \to \infty, Z \to \infty, G \to 0$, in other words, a drop-dead date! Obviously in the real world nothing is infinite; this is merely an indication that our simplistic mathematical model fails at this point. However, the period where this occurs may be a time of crisis with a major paradigm shift. If I knew how, I should disaggregate this case from all the other man-made hazards in the manner of Equation 8. But I do not know appropriate statistical weights, so I can only comment that my results may err on the side of optimism for survival.

Let us skip to the last solution, $\mu > 2$. Nothing physical prevents this case, but it seems remote from most of the world. Curiously, this case gives the cum-risk a jolt at

Table K. Cum-risk and haz-dev as functions of modern pop-time M.

Feedback range	Formula for technology U	Formula for cum-risk Z
$\mu < 1, \omega > 0$	$(\omega + 1)M^{\omega}$	$M^{\omega+1}$
$\mu = 1, \omega = \infty$	$\exp(cX)$	$\exp(cX) - 1$
$1 < \mu < 2, \omega < -1$	$\dfrac{(-\omega - 1)}{(L - M)^{-\omega}}$ Both $(-\omega)$ and $(-\omega - 1) > 0$ $U \to \infty$ as $M \to$ limit L	$\left(\dfrac{1}{L - M}\right)^{(-\omega-1)} - \left(\dfrac{1}{L}\right)^{(-\omega-1)}$ $(-\omega - 1)$ is positive Killer: $Z \to \infty$ as $M \to$ finite L
$\mu = 2, \omega = -1$	$\dfrac{1}{L - M}$ $U \to \infty$ as $M \to$ limit L	$\ln \dfrac{L}{L - M}$ Killer: $Z \to \infty$ as $M \to$ finite L
$\mu > 2, -1 < \omega < 0$	$\dfrac{\omega + 1}{(L - M)^{-\omega}}$ Note: $0 < \omega + 1 < 1$. $U \to \infty$ as $M \to$ limit L. Practically, analysis fails at a paradigm shift	$L^{\omega+1} - (L - M)^{\omega+1}$ Note: $0 < \omega + 1 < 1$. Survivable: Z stays finite as $M \to L$, but analysis stops due to paradigm shift; see U

a finite future time, but only a finite jolt. Evidently the system passes through crisis so quickly that exposure is minimal and survival feasible. Analysis cannot continue after the jolt because $U = \infty$ at that point, or in practical terms some cataclysmic paradigm shift violates our assumptions and completely changes the nature of the problem.

$$ \# \qquad \# \qquad \# $$

Let us apply the first case in Table K, the only one we develop, to find a formula for the survival predictor G_m, Equation K-1, also Equation 10. First let us split the cum-risk Z and the relative pop-time M into past and future. Then by the second of Equations K-9

$$ Z_p = M_p^{(\omega+1)} \qquad \text{and} \qquad Z_p + Z_f = (M_p + M_f)^{(\omega+1)} \qquad \text{(K-10)} $$

Substitution into Equation 10 (first equality), also K-1, evaluates GSP for man-made risks:

$$ G_m = (1 + M_f/M_p)^{-(\omega+1)q} \qquad \text{(K-11)} $$

$$ \# \qquad \# \qquad \# $$

Finally, let us restore the factor for natural hazards, $G_n = 1/(1 + T_f/T_p)$ from Section 4.1, to display the complete predictor, $G(f \mid p) = G_n^{1-q} \times G_m^q$, for the case in which the exponent q is known:

$$G(f \mid p, q) = G_n^{1-q} \cdot G_m^q = \left(\frac{1}{1 + T_f/T_p}\right)^{1-q} \cdot \left(\frac{1}{1 + M_f/M_p}\right)^{q(\omega+1)} \qquad \text{(below Eq. 14)}$$

The next to last paragraph in Section 3.1 explains that the negative exponents must sum to 1.0. In Section 3.3 statistics of London stage productions confirmed this. Here the exponents sum to $1 + q\omega$. This apparent discrepancy happens because M is not the cum-risk. The real cum-risk is Z in Equation 10, and in this case the exponents do sum to 1.0. It is merely a coincidence that the conversion from Z to M, Equation K-10, causes the second parenthesis in the equation above to look like a cum-risk factor.

Appendix L

Statistical weights for types of hazard

Sometimes we need a probability distribution that reflects no information about the process in question but merely lack of bias. Typically these are used as prior probabilities. For the interval 0 to 1, renowned scientist Harold Jeffreys favored

$$\text{uniform pdf:} \quad W(q) = 1, \tag{L-1}$$

and

$$\text{singular pdf:} \quad W(q) = \frac{1}{\pi\sqrt{q(1-q)}} \tag{L-2}$$

In Equation L-2, $1/\pi$ makes the total probability $= 1.0$. Note that this equation is singular (infinite) at both boundaries ($q = 0$ and 1). This seems appropriate for biological populations (humans included) because they are unstable: prone to boom or bust. (The classic example is lemmings, but I recall a day when I could not walk along the driveway in front of my house without stepping on baby frogs. This was in foothills, not a swamp.)

An example will help choose between L-1 and L-2. Suppose you know nothing about electrical conductivity. When you first encounter the subject, you might inquire about the relative conductivity of zinc and lead. In that case you would find them similar, zinc four times as much as lead, but that is not typical. If you inquire about aluminum and glass, you would find aluminum 3,000,000,000,000,000,000 times as conductive as glass, the opposite extreme. Similarly, consider an alien scientist arriving on Earth with no knowledge of our species. He would have no reason to think that q and $1 - q$ have comparable magnitudes, say $q = 0.3$ or 0.8. He would not be surprised if q is 0.007 or 0.9994. Hence, Equation L-2 with its two singularities makes more sense than Equation L-1 because they put heavy weight on extreme ratios between q and $1 - q$.

Unlike the alien we have a bit of knowledge and should modify Equation L-2 accordingly. First consider statistical weights for extinction; collapse of civilization comes later.

In Equation L-2 the singularity at $q = 0$ is appropriate, since zero implies no man-made threat. It represents the case in which human perpetrators lack the power or worldwide coverage to consummate extinction, which is entirely possible as discussed in Section 4.2. However, at $q = 1$, the singularity must change to a zero, $W(1) = 0$, because there is no chance that natural hazards vanish. If we keep the square root, the result is

$$W_{2\text{sp}}(q) = \frac{2}{\pi}\sqrt{\frac{1-q}{q}}; \qquad \text{mean: } \langle q \rangle_{2\text{sp}} = 1/4 = 0.25 \qquad (\text{L-3})$$

The subscript 2 denotes square root and sp denotes species survival. This formula is our basic uncontrived, statistically indifferent weight for the survivability estimates in Section 4.5. However, we should also look at a slightly contrived formula to test sensitivity, how much the final result deviates as a result of assumptions about statistical weight. An obvious choice is to substitute the cube root:

$$W_{3\text{sp}}(q) = \frac{3\sqrt{3}}{2\pi}\sqrt[3]{\frac{1-q}{q}}; \qquad \text{mean: } \langle q \rangle_{3\text{sp}} = 1/3 = 0.33 \qquad (\text{L-4})$$

This appears as the dashed curve in Figure 23. The reduced mean q makes this a more hazardous case, which is appropriate to offset biases in the less hazardous direction.

#   #   #

Finally we must modify $W(q)$ for the survivability of civilization. Recall that the two W_{sp} above have singularities at $q = 0$ because there may be no artificial hazard powerful enough to consummate complete extinction. However, to merely destroy civilization, a disaster need not reach the most remote places. A man-made hazard may suffice without nature's assistance. By simply removing the singularity from Equation L-3, we get the most obvious unbiased expressions:

$$W_{2\text{civ}} = \frac{3}{2}\sqrt{1-q}; \qquad \text{mean: } \langle q \rangle_{2\text{civ}} = 2/5 = 0.40 \qquad (\text{L-5})$$

$$W_{3\text{civ}} = \frac{4}{3}\sqrt[3]{1-q}; \qquad \text{mean: } \langle q \rangle_{3\text{civ}} = 3/7 = 0.43 \qquad (\text{L-6})$$

Appendix M

Extinction thwarted by civilization's collapse

Equation 14 in the text gives Gott's predictor for human survival since $G = G_m$ in our approximation. Recall that M denotes accrued population-time since about 1530 AD. Let us modify the notation to use fewer subscripts and make the future M_f look more like a variable of integration:

$$M_p \to M; \qquad M_f \to m$$

Then Equation 14 becomes

$$G = \left(\frac{M}{M+m}\right)^{(1+\omega)q} \tag{M-1}$$

Depending on q in the exponent, this applies either to civilization's collapse G_c or to species' extinction G_s. For the moment let us pretend that q is known. Later we can take the actual uncertainty into account by averaging as in Equation 16.

We want to know how survival depends on future *time*, but Equation M-1 tells us only how survival depends on future cum-risk m. To deduce m from time or vice versa, we use *projections* of future population, economy, technology, and such. These projections are more or less valid as long as things run smoothly. However, they break down completely if and when we suffer a cataclysmic event, such as collapse of its civilization. In the aftermath the world will be very safe owing to losses of population, economic activity, and technology. That is, $dm/dt \approx 0$. However, the projections have made no allowance for this event, we must devise our own. For use with projections let us define G_j as the survivability that takes into account the probability of civilization's collapse; see Section 4.4, Figure 26.

To find G_j, we start with the rate of species expiration, $-dG_s/dm$, and reduce it by the fraction of surviving species that are actually at risk, namely G_c/G_s. The remaining fraction has been rescued by the collapse that stops the threats. The adjusted expiration rate is

$$\frac{dG_j}{dm} = \frac{G_c}{G_s}\frac{dG_s}{dm} \tag{M-2}$$

First let us calculate the factor dG_s/G_s using Equation M-1 and the fact that this is the derivative of its logarithm:

$$\frac{dG_s}{G_s} = d(\ln G_s) = -(1+\omega)q \cdot d(\ln(M+m)) = \frac{-(1+\omega)q}{M+m}dm \tag{M-3}$$

To use Equation M-1 for G_c, it needs a new symbol p in its exponent to distinguish it from G_s:

$$G_c = \left(\frac{M}{M+m}\right)^{p(\omega+1)} \tag{M-4}$$

Note that $p > q$ because civilization's collapse cannot be slower than the species. Without species there is nothing left to collapse.

Substitution of Equations M-3 and M-4 into M-2 gives an expression for the rate of expiration dG_j/dm in terms of m, which one can integrate to find G_j, the desired result. One evaluates the constant of integration by requiring $G_j(0) = 1.0$. The result is

$$G_j = 1 - \frac{q}{p}\left[1 - \left(\frac{M}{M+m}\right)^{p(1+\omega)}\right] \tag{M-5}$$

Next, we must face the fact that we do not actually know p and q and must instead take a weighted average as in Equation 16. A vague probability density $W(p,q)$ expresses what little we do know. Apart from the restriction $p > q$, let us multiply separate probabilities as though p and q were independent. Let us therefore multiply the two preferred probability density functions $W_{2sp}(q)$ and $W_{2civ}(p)$ given by Eqns. K-3 and K-5 in Appendix K but renormalize them for the restriction $p > q$. The result is

$$W = \begin{cases} \dfrac{45}{32}\sqrt{\dfrac{(1-p)(1-q)}{q}}, & 0 < q < p < 1 \\ 0, & \text{otherwise} \end{cases} \tag{M-6}$$

The final result is the average of survivability G_j, Equation M-5 using M-6:

$$\langle G_j(m)\rangle = \int_{q<p}\int_0^1 W(p,q) \cdot G_j(p,q,m)\,dp\,dq \tag{M-7}$$

Putting $m = \infty$ in Equation M-5 gives the long-term asymptotic survival. The result is

$$\langle G_j(\infty)\rangle = 1 - \int_{q<p}\int_0^1 W(p,q)\frac{q}{p}dp\,dq \tag{M-8}$$

Curiously this is a single number, namely 70%, independent of both M and ω.

The initial expiration rate is especially important because it is about all we need to know for the next twenty years. Equations M-2 and M-3 are handy because $G_s = G_c = 1$ and $m = 0$. The result is

$$\text{Initial hazard rate} = -\left.\frac{dG_j}{dm}\right|_{m=0} = \frac{(1+\omega)}{M}q \qquad \text{(M-9)}$$

Appendix N

Initial hazard rates

To find the initial hazard rates for natural and artificial hazards, first rewrite Equation 15 in the form

$$G = \langle F \rangle_q \tag{N-1}$$

where $\langle \ldots \rangle$ denotes a weighted arithmetic mean, and F is given by

$$F = \left(\frac{T_p}{T_p + T_f} \right)^{1-q} \cdot \left(\frac{M_p}{M_p + M_f} \right)^{(\omega+1)q} \tag{N-2}$$

Next:

$$\ln F = (1 - q)[\ln T_p - \ln(T_p + T_f)] + q(\omega + 1)[\ln M_p - \ln(M_p + M_f)] \tag{N-3}$$

Let us introduce the prime notation for time derivatives, for example

$$X' \equiv dX/dT,$$

then:

$$\frac{d}{dT_f} \ln F = \frac{F'}{F} = \frac{q-1}{T_f + T_p} - \frac{q(\omega + 1)}{M_f + M_p} M'_f \tag{N-4}$$

Since $M = \int p \, dt$, $M'_f = p$. Evaluate Equation N-4 at $T_f = 0$ noting that $F(0) = 1$:

$$F'(0) = \frac{q-1}{T_p} - \frac{qp}{M_p}(\omega + 1) \tag{N-5}$$

Finally, take the average of Equation N-5 as indicated in Equation N-1 to find the initial hazard rate:

$$\Lambda = -G'(0) = -\langle F'(0) \rangle = \frac{1 - \langle q \rangle}{T_p} + \frac{\langle q \rangle p}{M_p}(\omega + 1) \tag{N-6}$$

The first term on the right pertains to natural hazards, the second to man-made. They may be separated; in particular the first term is negligible, and the second term is the expression reported in Section 4.6 in the main text, the initial hazard rate for man-made threats.

References

[1] Vinge, Vernor: "The Coming Technological Singularity: How to Survive in the Post-Human Era," *Whole Earth Review*, Winter 1993. Also, *VISION-21 Symposium*, NASA Lewis Research Center and Ohio Aerospace Institute (March 30–31, 1993).

[2] Rees, Sir Martin: *Our Final Century: Will the Human Race Survive the Twenty-first Century?*, Arrow Books Ltd., 2004.

[3] Joy, Bill: "Why the Future Doesn't Need Us," *Wired* magazine, cover story (April 2000).

[4] Simon, Julian: *The Ultimate Resource*, Vol. 2, Princeton U. Press, 1996.

[5] Crichton, Michael: "Let us Stop Scaring Ourselves," *PARADE* Magazine (December 5, 2004).

[6] J. Richard Gott, III: "Implications of the Copernican Principle for our Future Prospects," *Nature*, **363**, 315–319 (May 27, 1993).

[7] Thurber, James: *http://en.wikiquote.org/wiki/James_Thurber* (last accessed January 3, 2009).

[8] Webb, Stephen: *Where Is Everybody?* Praxis/Copernicus Books, Chichester, U.K./ New York, 2002.

[9] Brandon Carter: "The Anthropic Principle and its Implications for Biological Evolution," *Philosophical Transactions of the Royal Society of London*, **A310**, 347–363.

[10] "Doomsday Argument," Wikipedia, *http://en.wikipedia.org/wiki/Doomsday_argument* (last accessed January 3, 2009).

[11] Gott III, J. R., "Our Future in the Universe," in *Clusters, Lensing, and the Future of the Universe*, Trimble, V. and Reisnegger, A., eds. *ASP Conference Series*, **88** (1996).

[12] Leslie, John: *The End of the World: The Science and Ethics of Human Extinction*, Routledge, London, 1996.

[13] Mata, J. and Portugal, P.: "Patterns of Entry, Post-Entry Growth and Survival: A Comparison between Domestic and Foreign-Owned Firms," *Small Business Economics*, **22**, 283–298 (April 2004).

[14] Buch, P.: "Future Prospects Discussed," Scientific Correspondence, *Nature*, **368**, 107–108 (March 10, 1994).

[15] Sober, Elliott: "An Empirical Critique of Two Versions of the Doomsday Argument—Gott's Line and Leslie's Wedge," *Synthese*, **135**, 415–430 (2003). Also *http://philosophy.wisc.edu/sober/doomf.PDF*, p. 4 (accessed April 2009).

[16] Baldwin, J. R., *et al.*: Failure Rates for New Canadian Firms: New Perspectives on Entry and Exit," *Statistics Canada*, Minister of Industry, catalog # 61-526 (2000).

[17] Dunne, T., Roberts, M. J., and Samuelson, L.: "Patterns of Firm Entry and Exit in U.S. Manufacturing Industries," *The RAND J. of Economics*, **19**, 495–515 (Winter 1988).

[18] Goodman, Steven N.: "Future Prospects Discussed," Scientific Correspondence, *Nature*, **368**, 106 (March 10, 1994).

[19] Hoffman, Paul: *The Man Who Loved Only Numbers*, Hyperion, New York, 1998.

[20] Press, S. J. and Tanur, J. M.: *The Subjectivity of Scientists and the Bayesian Approach*, Wiley, New York, 2003, Sec. 5.4.

[21] Jeffreys, H.: *Theory of Probability*, Edition of 1939 or later, Clarendon Press, Oxford, Ch. 3.

[22] Audretsch, D.B., *et al.*: "New-Firm Survival and the Technological Regime," *Review of Economics and Statistics*, **73**, 441–450 (August 1991).

[23] Wearing, J. P.: *The London Stage 1890–1899: A Calendar of Plays and Players*, The Scarecrow Press, Metuchen, New Jersey, 1976. Six more volumes followed, one for each decade, the last being 1950–1959, Scarecrow Press, 1993.

[24] Zipf, G. K.: *Human Behavior and the Principle of Least Effort*, Addison-Wesley, 1949.

[25] Zipf, G. K.: *Selective Studies and the Principle of Relative Frequency in Language*, Harvard University Press, Cambridge, MA, 1932.

[26] Mandelbrot, B.: *The Fractal Geometry of Nature*, Freeman & Co, 1983.

[27] Google search of the Internet for "Zipf's law" plus a few other random sightings.

[28] Sober, Elliott: "An Empirical Critique of Two Versions of the Doomsday Argument—Gott's Line and Leslie's Wedge," *Synthese*, **135** 415–430 (2003). Also *http://philosophy.wisc.edu/sober/doomf.PDF* (accessed April 2009). See Note 5, p. 16.

[29] Van Valen, L.: "A New Evolutionary Law," *Evol. Theory*, **1**, 1–30 (1973).

[30] Ian McDougall, Francis H. Brown, and John G. Fleagle: "Stratigraphic placement and age of modern humans from Kibish, Ethiopia," *Nature*, **433**, 733–736 (February 17, 2005).

[31] Anders Johansen and Didier Sornette, "Finite-time singularity in the dynamics of the world population, economic and financial indices," *Physica A*, **294**, 465–502 (May 15, 2001).

[32] Price, Derek J. de Solla, *Little Science, Big Science*, Colombia U. Press, New York, 1963.

[33] U.S. Census Bureau, Historical Estimates of World Population, 8000 BC to 1950: *http://www.census.gov/ipc/www/worldhis.html* 1950 to 2050: *http://www.census.gov/ipc/www/idb/worldpop.html* (both accessed 3 January 2009).

[34] DeLong, J. Bradford, "Estimating World GDP, One Million B.C.—Present," *http://econ161.berkeley.edu/TCEH/1998_Draft/World_GDP/Estimating_World_GDP.html* (last accessed January 15, 2009). Lately this site has been online intermittently.

[35] U.S. Patent and Trademark Office: *http://www.uspto.gov/web/offices/ac/ido/oeip/taf/h_counts.pdf*—column 5 (last accessed January 3, 2009).

[36] *Nature* Magazine: *http://www.nature.com/nature/archive/index.html* (last accessed January 3, 2009).

[37] Vincent Larivière, Éric Archambault, and Yves Gingras: "Long-Term Variations in the Aging of Scientific Literature: From Exponential Growth to Steady-State Science (1900–2004)," *J. American Society for Information Science and Technology*, **59**(2) 288–296 (January 15, 2008). *http://www.ost.uqam.ca/Portals/0/docs/articles/2008/JASIST_Aging.pdf* (last accessed January 3, 2009).

[38] Zhivotovsky, L. A., Rosenberg, N. A., and Feldman, M. W.: "Features of Evolution and Expansion of Modern Humans, Inferred from Genomewide Microsatellite Markers," *American J. Human Genetics*, **72**, 1171–1186 (2003).

[39] John H. Moore: "Minimum viable human population," Dept. of Anthropology, University of Florida, *http://web.clas.ufl.edu/users/moojohn/moojohn@anthro.ufl.edu* (last accessed October 2008).

[40] Private meeting with Patrick Mock, former member of an Antarctic team.

[41] Powell, Corey S.: "20 Ways the World Could End Swept away," *Discover*, **21** (October 2000).

[42] Smith, Quentin: "Essay on Leslie's The End of the World," *Canadian J. of Philosophy*, **28**, 413–434 (1998).

[43] Pain, Stephanie: "Interview with JoAnn Burkholder," *New Scientist*, **166**(2241), 43–45 (June 3, 2000).

[44] Cloud, Preston: "The Biosphere," *Scientific American* (September 1970). *See also*: Environmental effects of ozone depletion: 1998 Assessment *http://www.gcrio.org/UNEP1998/UNEP98p49.html#T6 UNITED NATIONS ENVIRONMENT PROGRAMME* (last accessed January 3, 2009).

[45] Trace Gas Exchange at *http://www.gcrio.org/UNEP1998/UNEP98p49.html* (last accessed January 5, 2009).

[46] Arens, N.C. and West, I.D.: "Press/Pulse: A General Theory of Mass Extinction?" GSA conference paper, 2006.

[47] Posner, R.: "The End Is Near," *The New Republic* (September 22, 2003), p. 34. Book review of Atwood, M.: *Oryx and Crake*, Doubleday.

[48] J. F. Kasting, Owen B. Toon, and James B. Pollack: "How Climate Evolved on Terrestrial Planets," *Scientific American* (February 1988).

[49] Kunzig, R.: "20,000 Microbes Under the Sea," *Discover* (March 2004), pp. 34–42.

[50] Goldberg, D.: "John du Pont Found Guilty, Mentally Ill," *Washington Post* (February 26, 1997), p. A01.

[51] Many sources, for example Simms, Andrew: *http://www.independent.co.uk/opinion/commentators/andrew-simms-we-should-introduce-a-maximum-wage-674834.html* (last accessed January 3, 2009).

[52] Kurzweil, R.: *The Age of Spiritual Machines*, Viking, New York, 1999.

[53] Kruglinski, S.: "Interview: Gerald Edelman," *Discover* (February 2009), p. 64.

[54] Meadows, D. H., *et al.*: *Limits to Growth*, Potomac Associates, New York, 1972. Update: Meadows, D. L., *et al.*: *Limits to Growth: The 30-year Update*, Chelsea Green Publishing Co., 2004.

[55] Stommel, H. and Stommel, E.: "The Year without a Summer," *Scientific American* (June 1979), p. 176.

[56] R. L. Schweickart, *et al.*: "The Asteroid Tugboat," *Scientific American* (November 2003), p. 54.

[57] Lists of philanthropists: *http://images.businessweek.com/ss/08/11/1124_biggest_givers/1.htm* and *http://www.businesspundit.com/25-billionaires-and-millionaires-that-became-philanthropists/* (both accessed January 3, 2009).

[58] "The Limits to Growth revisited," *Quantum: The Magazine of Math and Science*, special issue (September/October 1997).

[59] Crichton, Michael: *Rising Sun*, Alfred A. Knopf, 1992.

[60] Crichton, Michael: *State of Fear*, Harper Collins, 2004.

[61] Wattenberg, Ben J.: *Fewer: How the New Demography of Depopulation Will Shape Our Future*, Ivan R. Dee, Publisher (2004).

[62] Nonaka, K., Minura, T., and Peter, K., "Recent fertility decline in Dariusleut Hutterites: An extension of Eaton and Mayer's Hutterite fertility study," *Human Biology*, **66**, 411–420 (June 1994).

[63] Hardin, G.: "The Tragedy of the Commons," *Science*, **162**, 1243–1248, 1968.

[64] The Quotations Page: *http://www.quotationspage.com/quotes/Joseph_Stalin/* (last accessed January 3, 2009).

[65] Herman E. Daly: "Economics in a Full World," *Scientific American* (September 2005), p. 100.

[66] Partha Dasgupta: *Human Well-Being and the Natural Environment*, Oxford U. Press, New York, 2001.

[67] Boulding, K. E.: "The Economics of the Coming Spaceship Earth," in H. Jarrett (ed.), *Environmental Quality in a Growing Economy*, Johns Hopkins U. Press, Baltimore, 1966, pp. 3–14.

[68] Leslie, John, *The End of the World*, Routledge, London, 1996, Ch. 3.

[69] Moravec, H., *Mind Children: The Future of Robot and Human Intelligence*, Harvard U. Press, Cambridge, MA, 1989.

[70] John M. Edmond and Karen Von Damm: "Hot Springs on the Ocean Floor," *Scientific American* (April 1973), p. 78.

[71] Miller, S. L. and Cooper, C.: "The Aquarius Underwater Laboratory: America's 'Inner Space' Station," *Marine Technology Society J.*, **34**, 69–74 (Winter 2000).

[72] O'Neill, G.: *The High Frontier*, William Morrow, New York, 1977.

[73] Diamond, Jared: *Collapse: How Societies Choose to Fail or Succeed*, Viking Penguin, New York, 2005.

[74] Gott, J. R.: "A Grim Reckoning," *New Scientist* (November 15, 1997).

[75] Gott, J. R.: "Longevity of the Human Spaceflight Program," in: *New Trends in Astrodynamics and Applications III*, American Institute of Physics Conference Proceedings, Volume 886, Edward Belbruno, editor, AIP, New York., 2007.

[76] Gott, J. R.: "The Chances Are Good You Are Random," *New York Times* (July 27, 1993).

[77] Jaynes, E. T.: *Probability Theory: The Logic of Science*, Cambridge University Press, Cambridge, U.K., 2003.

Index

(**bold** numerals denote a principal discussion; *italics* denote figures, pictures, and tables)

Accuracy, univariate better than bivariate, 68
Adaptation, 121
Aftermath, safety, 5, 68, 87, 93
AIDS, 114–115
Altruism, misguided, 121
Analytic models, complementary to simulation, 111
Anomalous mortality of stage productions, 47–48
Antarctica, 93–94
Approaches to survivability, four, 6, **14**
Asteroid, 57, 59, 82, 96, 113, **126**
Audience response, 122–123

Bayes' theory, 39–41, **139**
  modified for partial knowledge of Q, 142
Behavior of mean future as sample size grows, 23
Behavior, simulation of human, 110
Benford's law, **33**, 111
Biosphere 2, 98–99
Bivariate example, 62–64
Black death, 95
Blackout, 101
Body counts for separate hazards, 163
Bolide, 5, 82, 98–101, 113–114
Boulding, Kenneth Ewart, 121
BPC (billions of people centuries), 6
Business firms, survival of, **41–45**

Butterfly in Brazil, tornado, 4

Carbon dioxide, 68–9, 87, **95**, 110
Chaos (chaotic), **4**, 29, 36, 79–80, 117
Chapters, capsule summary, 9
Civilization, collapse, 5
  age and survivability, 84
Clinton and Bush, 115
Coincidence, 99–100
Colonize solar system/galaxy, 6
Complexity theory, 4
Composition, production, or run, 46
Confidence, degree of, 6
  threshold = 90%, 22
Confirmation, Audretsch's empirical, 45
Conscious artifact, 109
Conspiracy, 103
Crazy Horse, 115
Crichton, Michael, 2, 112, **117**
Cumulative risk (cum-risk)
  in general, **20–21**, 32, **34–37**, 187–188, 189–190
  to humanity, man-made hazards, 54, 57–81, 193–195
Curiosity, 121

Deccan Traps in India, 99
Deforestation in Amazonia, 107
DeLong, J. Bradford, 74

Development, measure of hazardous, **72–78**
Die of unknown provenance, Guy's 25
Die, loaded, 15
Dinoflagellates, 98
Dinosaurs, **5, 99–100**, 113, **121**
Disaggregation, **58–64**, 74, 92, 194
   mortality, 163–165
Dispassionate look at social values, 70
Doomsday Argument, **7–8**, 32
Doomsday vault, 2
Du Pont, John, 105
Dual hazards, performances and time, 167
Duration given, age unknown, 40–1

Earth-like planets, 5, 67
Easter Island (Rapa Nui), 126–127
Economics, neoclassical, 121
Electric stork, 102
Exploration, age of, 78–79
Exponent sum $q + r$, substantiating, 62
Exponential decay, *14–15*
Exponents for relative importance, 58
Exposure to hazards, 4, 57
Extinction insurance, 5, **122–123**
Extinction rates of prehistoric taxa, 159
Extinction thwarted by civilization's
   collapse, 199

Fermi's paradox, 5, 125, 127
Fertility, 118
Ficticium, radioactive, 15
Future research, 50, **65–66**

Galaxy with humanoid species, 66–68
Gates, Bill and Melinda, 116
Gauge, virtual risk, 57
Genetic algorithms, 108
Geometric average, 60
Gestation period J, **12–13**, 58
Goodman, Steven, 28–29
Gott III, J. Richard, 4, **8**
Gott's predictor, 21–22, **30–31**
Grand Galactic Book of Knowledge, 67, 87
Greenhouse, 101
Gross world product, 73, 78
Guy and his die, 25

Half-life, 6
Hawking, Stephen, 2
Hazard rates, current, 90–91
Hazardous development (haz-dev), 73
Hazards during 1970 compared to 2005,
   80–81
Hiatuses, startup costs, 62
Homogenization, 74, 81, 100, 106, **121**
Hostile androids, 109
Human race, survivability of, 86
Human survivability, formulation, 69–81
   defining cum-risk Z, 71–72
Humanoids in our galaxy, 66–68, 120

Illegal travel, 95
Imagination, simulation with, 111
Impresario, bivariate formula, 58, 82
Independence, statistical, 59
India and China, 105
Infinite mean duration, 24, 137
Initial hazard rates, 203
Instability, 101
Isotope production, 16, 17

J, gestation period, **12–13**
Jeffreys, Sir Harold, 37, 84
Jorj, oil shortage, 34–35
Joy, Bill, 2, 109
Jupiter, collision with Shoemaker–Levy, 114

Land mines, 106–107
Laplace, Pierre Simon, 39
Latent killer, 100
Leslie, John, **8**, 96
Little Ice Age, 112
Logarithmic scales, 42
Logic diagram, part 1, **39–40**
   part 2, figure 18, **64–65**
Logic outline of the overall plan, 187

Main-sequence shows, 62
Man-made hazards, cum-risks for, 193
McMurdo Station, 94
Mean duration, infinite, 24, 137
Median, 7, **13**
Mental disconnect, 122–123
Microcosms for humanity, 55
Microcosms, 3

Moore, Gordon, 106, 107, 120
Moore's law, 107–108
Moravec, Hans, 125
Multiple hazards, 189
Murphy's tavern, 28, 29, 31
Mutant phytoplankton, 97–98

Native copper, 124
Near misses as proxies for hits, 82
Noninformative prior for age and duration,
    145

O'Neill's cylinder, 126
Obsolescence, residual 24, 38, 42, 50, 91
Occasional shows, 62
Offense, huge advantage over defense, 95
Oil shortage, Planet Qwimp, 34–35
Oldest stage show, oldest business, 24
Opening in specific year, 46, 49
Optimistic, biases make analysis too
    explosive technological growth, 91–92,
        194
    loss of vitality, obsolescence, 48–49, 92
Overall plan for survivability calculations,
    187
Overpopulation limiting survival time, 86,
    90
Overrated natural hazards, 112–114

Paradox of Jays and Kays, 57–58
Paradox of no natural hazards, 82
Paradox of vanishing gestation period,
    30–31
Pascal, Blaise, 39
Patents, U.S., 76
Perils ranked emotionally, 119
Pharmacology, 100
Philanthropy, 115–116
Playing on specific dates, 46, 49
Population, world, whom to count, **71–72**,
    73
Population-time (pop-time), 6, **74**
Portuguese business firms, 11
Posterior probability, **21**, 54
Posterior survivability, 135
Preindustrial, 68
Preservation of historic structures, 66,
    189–190

Press/pulse, 99
Price, Derek, 73, 91
Principle of indifference, **25–27**, 34
    restoration, 37, 72
Principle of insufficient reason, **25–27**, 34
Prior probability, **21**, 54
Privacy in survival habitat, 126
Probabilities, weighted average, 19
Probability theory, review, 18
Product rule, AND, **18**, 57–58, 60, 63
Projection of population-time, 88

Q, fraction surviving beyond T, 11
Quacksilver, 29

Radioisotopes, 14
Rank, statistical, 52
Reaction, public to simulation, 111–112
Red tide, 98, *back cover*
Rees, Sir Martin, 2, 95, 117, 126
References, 205
Robotic colonists, 127
Robotics, 107
Robots, self-sufficient, self-replicating,
    102–103

Safer world, prospects, 117
Scatter diagram for London Shows, 62–63
Scenarios for extinction, 97
Second chance? 123–125
Secret Eugenics Society, 103
Self-extinction, 5–6, 32, 68–71, 79, 82–84,
    120
Self-sufficient colonies at harsh locations
    on Earth, 126
Shooting gallery, Zyxx, 35–36
Singularity, 1, 5, 91
Social problems, hazardous technical
    solutions, 122
Spaceman Jorj, oil shortage, 34–35
Space-traveler Zyxx, shooting gallery,
    35–36
Spanish influenza, 45, 95
Stage productions running on specified
    dates, 147
Stage productions with dual cum-risks, 167
Stage productions, survival of, **45–51**
    bivariate: duration, performances, **62–64**

Statistical weights $W(q)$ for hazards to humanity, 83–84, 197
Statistician Stacy, 28, 29, 31
Statistics of leading digits, 33
Substantiation, strong or weak, **61**, 66
Sum rule, OR, 18, 83
Sunspots, 112
Survivability of entities of unknown vulnerability, 132
Survivability, prior for varying hazard rate, 131
Survivability, prior from unknown hazard rates, 129
Survival colonies, 93
Survival depending on projected pop-time, 89
Survival habitat, 98, 125
Svalbard Global Seed Vault, 2

Taxa, prehistoric, 54
Ted Turner, 116
Timelines for Gott's survival predictor, 27
Tipping point, 4, 74, 111, 118
Triage, 114
Tropics, nature take back, 106–107

Ultimate predictor assuming $T_f \ll T_p$, 85
Uncertainties, hierarchy, 17–18
Underwrite insurance, 91

Unknown hazard rate, prior survivability, 129
Utility meters, 20, 69

Vague prior probability of age and duration, 145
Van Valen, Leigh, 54
Varying hazard rate, prior survivability, 131
Venn diagram, 116
Venus, 101
Viewpoint, necessity of detached, 68–69
Viewpoints, four, 6, **14**
Vinge, Vernor, 1
Vitality, loss of, 24, 38, 43, 50, 91
Volcanic Deccan Traps, 99
Von Neumann, 1

War, insignificance for survival, 97
Wealth, private, 105–107
Wearing, J. P., 45, 49
Webb, Stephen, 5
Wild Cards, 105
World simulation, 110–112, 116
World Wars, 50, 62

Zipf, George Kingsley, 52–53
Zyxx, shooting gallery, 35–36